# FROM THE COTTON FIELD TO THE COTTON MILL

# FROM THE COTTON FIELD
# TO THE COTTON MILL

A STUDY OF
THE INDUSTRIAL TRANSITION
IN NORTH CAROLINA

BY

HOLLAND THOMPSON

  BOOKS FOR LIBRARIES PRESS
FREEPORT, NEW YORK

First Published 1906
Reprinted 1971

HD9877
N8T5

INTERNATIONAL STANDARD BOOK NUMBER:
0-8369-5663-X

LIBRARY OF CONGRESS CATALOG CARD NUMBER:
71-148900

PRINTED IN THE UNITED STATES OF AMERICA

# PREFACE

THE author has spent the greater part of his life in the section described. While living in a rapidly growing mill town ten years ago, the sight of scores of wagons transferring scanty household goods from farmhouses to factory tenements awakened his interest in the sudden transformation of farmers into factory operatives.

His interest in the problem has cost much time and trouble. He has read everything available upon the subject, has sifted and compared dozens of statistical tables, and has compiled others. He has visited many mills, has talked with dozens of mill owners, managers, superintendents, overseers, and operatives. The children in the mill, at school or upon the streets, and the parents at home have not been overlooked. The teachers, ministers, and church workers in the mill villages have helped. The business men, the officers of the law, the farmers, and the laborers, black and white, all have added something.

Removal from the state gave the opportunity of visiting similar manufacturing establishments in other states, and has also afforded perhaps a truer perspective. However, a part of every year has been spent in North Carolina, and impressions and opinions have been tested by time, the great touchstone of truth.

Greater hesitation in delivering final judgments has followed increasing knowledge. The interpretation of the life of a people is no slight undertaking. The author cannot speak so confidently as he would have done five years ago. Many phenomena, apparently permanent, have proved to be transient, and unexpected elements have increased the complications. At least he has written the truth as the truth appears after studying the problem for ten years.

While the study has been confined to North Carolina, much is equally applicable to other Southern states. Repetition has been unavoidable, since different phases of the problem have been taken up in turn. Every effort to eliminate the unessential has been made, however, and many paragraphs might easily be extended into chapters.

The list of those who have given assistance is so long that separate credit is impossible. Especial thanks are due to F. L. Robbins, Esq., of Salisbury, North Carolina. His knowledge and experience guarantee the correctness of the technical chapters, and his sympathy and insight have been valuable.

HOLLAND THOMPSON.

Townsend Harris Hall,
    April, 1906.

# CONTENTS

| CHAPTER | | PAGE |
|---|---|---|
| I. | THE PROBLEM . . . . . . . | 1 |
| II. | THE STATE AND ITS PEOPLE . . . . | 16 |
| III. | DOMESTIC MANUFACTURES AND THE BEGINNING OF THE TEXTILE INDUSTRY . . | 37 |
| IV. | THE GROWTH SINCE 1861 . . . . | 55 |
| V. | THE PRESENT STATE OF THE INDUSTRY . | 74 |
| VI. | THE REAL FACTORY OPERATIVE . . . | 96 |
| VII. | THE OPERATIVES AT WORK . . . . | 118 |
| VIII. | WAGES AND COST OF LIVING . . . . | 137 |
| IX. | SOCIAL LIFE AND AGENCIES FOR SOCIAL BETTERMENT . . . . . . . | 162 |
| X. | THE DEVELOPMENT OF A CLASS CONSCIOUSNESS . . . . . . . . | 182 |
| XI. | THE RELATIONS OF EMPLOYER AND EMPLOYED . . . . . . . . | 200 |
| XII. | THE CHILD IN THE MILL . . . . | 219 |
| XIII. | THE NEGRO AS A COMPETITOR . . . | 248 |
| XIV. | CONCLUSIONS . . . . . . . | 269 |

# FROM THE COTTON FIELD TO THE COTTON MILL

## CHAPTER I

### THE PROBLEM

WHEN an old state — one of the original thirteen — builds almost two hundred cotton mills within twenty years, and also enters largely into other manufactures, evidently a great economic change is indicated. The fact that the capital has come chiefly from a multitude of small investors within the state, makes the change more striking. When, with almost imperceptible immigration, from 150,000 to 200,000 persons are transferred from the country — perhaps from the very farms — where they and their ancestors have lived for more than a century, to live in towns or factory villages, and receive their pay in wages rather than in commodities, the social changes must be equally important.

North Carolina has been and is yet a rural state. No city has ever dominated, or even influenced, any considerable portion of the territory. In 1900 there was not a single city with a population of 25,000. There were only six towns with more than 10,000, and only twenty-eight with more than 2500. Of a total population of 1,893,810, only 17.9 per cent. lived in incorporated towns at all, no matter how small, compared with 47.1 per cent. in the United States as a whole. Only Alabama, Mississippi, and Arkansas of the Southern states showed a smaller proportion of town dwellers. Only 12.1 per cent. gained a livelihood by manufacturing or mechanical pursuits, while 64.1 per cent. were employed in agriculture or fisheries. But these figures differ decidedly from those of 1890. Then only 13 per cent. lived in towns, 9.6 per cent. were engaged in manufacturing, while 69 per cent. were engaged in agriculture. Since 1900 the percentage of those engaged in manufacturing has steadily increased, and the transition to an industrial society is well begun. The state stands third in the manufacture of cotton; the product of the cotton-seed oil mills is important; North Carolina furniture is shipped to South America and South Africa; and North Carolina tobacco is sold over the world.

The state is being influenced profoundly by the transfer of a population by families instead of by individuals from the country to the town. Now, between an agricultural and an industrial population are many points of difference. The manner of life is unlike; the opinions are generally opposed; the ideals are not the same. As yet the division line in North Carolina is not sharp and clear. There is no manufacturing section in which agriculture is merely subsidiary. Cotton mills are located in more than half the counties of the state, and other industries are more or less scattered. There is no sharply defined operative class, for the workers in the mills and factories of North Carolina were either born on the farms, or are only one generation removed, and the tang of the soil still persists. With the making of operatives and artisans from farmers we have to deal.

Yesterday the mill operatives produced raw material for others to fashion; to-day they fashion it themselves. They were landowners or at least land renters with all the rural independence. Now they work at the overseer's nod, and receive their pay in wages rather than in products of the soil which they have directly created. Instead of living remote from neighbors, they are crowded into factory villages where they may talk from house to house. They spend the larger part of their waking time within walls, tending complicated machines instead of working in the open air with a few simple tools. In the country the work was irregular and an occasional holiday might be taken without apparent loss. In the mills loss of wages and the displeasure of the overseer follow any departure from absolute regularity. The operative must work every day and the whole of the day.

Such a radical change in manner of life must affect them physically and mentally. They must learn how to live in towns, to adapt themselves to their surroundings. The children worked on the farms, as they have done since farming began, but here they are subjected to constant instead of intermittent demands upon their strength and endurance. The mental activity of all must be influenced; a quickening or a deadening must follow.

Their social, religious, and political ideas are undergoing change. The gregarious instinct develops rapidly, and solitude, once no hardship, becomes unendurable. The religious ideas and organization which served the rural inhabitant seem not so satisfactory to the factory worker. The church is becoming alarmed to find that it is losing its hold upon the factory population. Political unrest is not yet general, but in a few localities the workers are slowly becoming conscious of themselves. Feeble attempts to organize a Socialist propaganda may be seen. The labor agitator is at work.

Those left on the farms are affected by the withdrawal of population, a part of which goes to the towns for employment in the various industries, and another part to invest its capital in trade or manufacturing rather than in agriculture. Both the churches and the schools feel the loss. Neighborhoods once attractive from a social standpoint are now lonely. On the other hand, the establishment of little towns in the fields and woods around the widely distributed mills affords new markets for farm produce. The wages for farm labor — for a long time either stationary or decreasing — rise because of the increased demand and the smaller supply, and improved machinery and more intensive farming are necessarily introduced. The rural telephone and improved roads, — both largely the results of the increased commercial and industrial activity, — together with rural mail delivery, help to bring the country communities into closer touch with the outside world.

The negro also is directly affected. The increased population and activity in the towns make opportunities for a larger number as servants or as laborers. Lumbering, or railway construction and improvement, have drawn away others from the farms. Those who remain receive larger wages, or may rent better farms than was possible before. The greater demand for their labor brings about greater consideration and greater intolerance. Less and less patience is exhibited toward the worthless and the indolent. At the same time, the faithful and reliable tenant or laborer receives increasing kindness and consideration.

The increase of population and wealth in the old towns is working many changes. Communities which had altered little since the days of Cornwallis are feeling the modern industrial spirit. "Business" is being exalted to a position heretofore unknown. A type of shrewd, calculating, far-sighted business man is being developed. The "Southern Yankees" devote themselves exclusively to their work and need ask no favors in any contest of commercial strategy. Social lines are shifting. Families which have decidedly influenced the spirit of the community become less prominent, unless they take part in the new movement. There are signs of class distinctions based upon wealth and business success.

The whole attitude of mind has changed more during the last fifteen years than in the fifty preceding. The Civil War did little more than to intensify the convictions previously existing. That acute, though often unfair, critic of Southern life, Judge Tourgée, well says, "It modified the form of society in the South but not its essential attributes." Reconstruction fixed these convictions more firmly. Now old prejudices and fixed ideas, political and social, show signs of weakening. Independent voting is no longer uncommon. Only the prominence of the race question prevented a greater division upon national lines in 1904. A military record no longer outweighs all other considerations. Not a single member of the present Congress from the state was a Confederate soldier. Commercialism is doing what bayonets could not do.

The ideal of success is changing. An increasingly large proportion of the college graduates adopt a business career, or go into the mills and factories to learn every process in spite of the dust and the grime. The state is growing more like industrial societies everywhere. Agricultural societies may show much variation, but industrial communities tend more toward a type. Nevertheless the influence of the old civilization is felt through the expression of the new, and modifies it almost in every detail.

These are the phenomena with which we have to deal. The task of this paper is to sketch some of these changes while in process; to show how this new industrialism suddenly introduced is affecting the economic and social life of the people. To make a section through the state and study all the kaleidoscopic relations would require many volumes. The most that can be attempted is to give a general view, and then to show in some detail the life of the thousands, suddenly transferred from agricultural to industrial employment, particularly in cotton, and to study how they are adjusting themselves to their new environment. An honest effort to state, calmly and dispassionately, actual not fanciful conditions will be made. Only incidentally will there be any attempt to predict the future other than to point out tendencies. That must wait for more complete studies. Few similar investigations have ever been made, and they have dealt, principally, with the growth of capital and the rise of the entrepreneur, rather than with the development of an operative class.

Studies of other countries or of other sections have not dealt with precisely the same phenomena. Before the industrial revolution, England was already a manufacturing country through the thousands of hand looms in the weavers' cottages. The factory system first brought these operatives together and furnished power. The first effect was to drive those unable to find a place in the new system back to the soil already crowded, or to throw them upon the parish. In North Carolina increased opportunities for profitable employment in every line of industry have followed the change.

The transformation in this state is more nearly like that in New England seventy-five years ago, but still with decided differences. In 1810, according to Tench Coxe, the value of the textile products of North Carolina in the domestic system was greater than that of Massachusetts produced by hand plus that of the few factories then existing.[1]

[1] Statement of Arts and Manufactures in the United States, 1810: N. C., $2,989,140; Mass., $2,219,279.

North Carolina, induced by considerations which will be discussed hereafter, turned her activity into other channels, and as a manufacturing state grew less important for half a century. Massachusetts, aided by the policy of the general government, continued to develop along industrial lines. Every improvement in method was adopted. Increasingly expensive and complex machines, often bought from the profits from simpler types, were installed. The plants grew in size and cost, and the amount and proportion of capital invested in manufacturing greatly increased.

When North Carolina again entered the contest for industrial success, the conditions of the problem were different. The industry was firmly established in another section, which had the prestige of long-continued success, controlled all the channels of the trade, and had a great body of skilled operatives. Greater capital was required and competition was keener. The state had almost lost the traditions of an industrial past and that difficult task, the removal of an agricultural population by families rather than by individuals, was to be accomplished. Moreover, the influence of the presence of the negro must not be underestimated.

Another factor of the difference which has its place, and must not be neglected, even in a purely economic study, is the essential difference between Northern and Southern character and attitude of mind, — a difference distinct from any question of an aristocratic structure of society. What has produced these differences need not be discussed here. The differences exist. The North and the South are two countries with different ideals, different prejudices, different standards. France and Germany are no more unlike than some portions of the United States. As a whole, the differences are certainly as great as those between England and Ireland. Any attempt to form comparisons and judgments without taking into consideration these ingrained differences will be of slight value.

Little aid in this study can be had from statistics. Complete and accurate figures concerning that portion which may be reduced to tables cannot be procured. The Census Reports do not give those facts which are of most interest and value. The Bulletins of the Department of Commerce and Labor are not broad enough in their scope. The State Bureau of Labor and Printing has no power to require answers to its inquiries, and its funds are so small that canvassers cannot be employed. Voluntary answers are the sole dependence. The questions necessarily are vague and general, and even these are often inaccurately answered, or are not answered at all.[1] Manifestly no private individual can gather full statistics.

But if figures, accurate at the time of collection, were secured, they would be obsolete almost by the time they were printed. The

[1] Letter from former commissioner.

state of the cotton industry, with which we shall deal particularly, is dynamic in the extreme. New mills are completed, old ones are enlarged, product is diversified, new machinery is introduced making new wage scales, night work is begun or discontinued. The surroundings of the little mill in the country with a few hundred spindles, and of the large establishment in a mill center, vary greatly, and the operatives move from one to the other with frequency. Further, the life of a people cannot be put into columns and averaged. That noted statistician, the lamented Professor Mayo-Smith, in fact declared that the opinions of trained observers were worth more than statistics, in estimating the relative welfare of different communities.

Yet underneath all the diversity there are constant factors, tendencies strongly marked, which may be described and analyzed if one studies the people as well as the material facts. We may be able to say "how" even if we cannot say "how much."

## CHAPTER II

### THE STATE AND ITS PEOPLE

THOUGH the first attempts to plant English settlements within the limits of the United States were made upon North Carolina soil, following the exploring expeditions sent out by Sir Walter Ralegh in 1584, their failure left the country long unoccupied. Meanwhile the Virginia settlements were spreading. Soon after 1650 straggling pioneers, induced by the desire for the rich land along the eastern streams, began to come within the present limits of the state. The old belief that these early settlers were driven by religious persecution to seek new homes, seems to have little foundation.[1]

In 1663 a charter for Carolina was issued

[1] See the careful researches of Weeks, "Southern Quakers and Slavery," *J. H. U. Studies*, extra volume, xv.

by Charles II to eight Lords Proprietors. Their territory was extended and their power confirmed two years later. All the privileges and powers pertaining to the bishopric of Durham were granted, and the Proprietors attempted to form a province modeled upon the County Palatine of Durham. The celebrated Fundamental Constitutions contemplated the establishment of a European feudal system, with orders of nobility, commons, and slaves, in a new country so thinly settled that it can hardly be said to have been settled at all.

The story of the failure is long and need not be told in detail. When the Crown again took control in 1729, the province had been divided into North and South Carolina and settlements had been made along the sounds and streams of the eastern section. Englishmen from Virginia, from Barbados, settlers direct from the mother country, German Palatines, Swiss, French Huguenots, and a few New Englanders made up a population amounting possibly to 30,000 whites and 6000 negroes.[1] Some large tracts of land had been granted, and there was already the beginning of a plantation system which increased in importance with the years. With all the mixture of nationality, however, English ideas and ideals were dominant.

Estates were never so large as in Virginia or in South Carolina. With a few exceptions, 660 acres was the largest grant made by the Proprietors, and 640 was more common. This policy was in striking contrast to their course in South Carolina. When the Crown assumed control, the policy of small grants was continued, though a few large tracts were granted for speculative purposes.[2] As a result, North Carolina became a province of small planters and farmers, compared with her neighbors.

While such settlers were filling up the East, and adventurous individuals were making their way up the streams toward the West,

[1] Raper, "North Carolina" (1904).
[2] *Ibid.*, pp. 108, 109, 118.

pioneers of another type occupied that section. Soon after 1830 wagon trains from Pennsylvania came down through the Shenandoah valley and settled upon the broad stretch of territory included in the valleys of the Catawba and the Yadkin. These were immigrants or the children of immigrants from the North of Ireland, the so-called "Scotch-Irish" who did so much to subdue western Pennsylvania.

When they began to feel crowded there, the overflow followed the foothills of the Alleghanies, and the western sections of Virginia, North Carolina, South Carolina, and western Georgia received a valuable population. Others landed at Charleston instead of Philadelphia, followed the rivers toward the northwest and met the southern current along the Yadkin. Stern, adventurous, religious, they made ideal pioneers, and from them developed that sturdy, independent middle class which helped to give North Carolina its peculiar characteristics.

Their occupancy of the wide territory was soon shared by Germans, also coming chiefly from Pennsylvania, though some landed at Charleston and joined their countrymen in what are now the counties of Davidson, Rowan, Cabarrus, Mecklenburg, Gaston, Lincoln, Catawba, and Iredell. About 1750 a colony of the Unitas Fratrum, better known as Moravians, bought a large tract of land around the present town of Winston-Salem. This they held in common, and urged on by religious zeal made great improvements. Small colonies from the back counties of Virginia and from Maryland also made settlements along the Yadkin, and a strong Quaker influx occupied the present counties of Randolph, Chatham, Alamance, Surry, and Guilford.

After the battle of Culloden in 1746, Scotch Highlanders came to Wilmington and ascended the Cape Fear River to Cross Creek, the present town of Fayetteville. From this nucleus they spread over a half dozen counties, where few except Scotch names are heard to-day.

During the Civil War, whole companies of "Macs" were enlisted. A colony of Irish direct from Ulster had been planted in the neighboring county of Duplin in 1730.[1]

Further, in few cases did these different nationalities locate upon the same territory where association might wear away characteristic peculiarities. The "consciousness of kind" was strong enough to segregate those of the same language, religion, and habit of mind. There was little communication between the different settlements, and definite characteristics of the eighteenth century persisted until late in the nineteenth.

In one county the distinction between the German and the Scotch-Irish has disappeared only within the twenty-five years just past. The jealousy between the "Dutch" and the "Irish" side was strong, and there was little association and less intermarriage. Idioms and expressions heard frequently in one part of the county were hardly intelligible in the

[1] Hanna, "The Scotch-Irish" (1902).

other. Culinary processes were different. The location of the courthouse, or the struggle for county officers, might be the occasion for a bitter contest.

If there was little communication between the different neighborhoods, there was less between the sections of the state. In the East were large, rich plantations upon the sounds and confluent creeks and rivers. Communication here was easy, as few points are more than five miles from water. Along with the large landowners were individuals to whom the much-abused term "poor whites" might be applied with more or less accuracy. Fish, oysters, wild fowl, and game were plentiful; the land was rich, and the procuring of a bare subsistence was too easy to require much work. To-day it is estimated that two months' work in every year will enable a family to live with some degree of comfort.

During the colonial period, English ideas governed this section as they did in Virginia. Even to-day those counties north of the

Roanoke River belong to Virginia rather than to North Carolina. Pieces of old English furniture and bits of English china and silver are to-day treasured heirlooms.

Back of the rich alluvial lands were the pine barrens to which the thriftless were gradually driven by the inexorable working of economic law. Until, with the development of the refrigerator car, it was discovered that this region would bring large returns in the trucking industry, the land was of little value except for the pine forests, which have been ruthlessly destroyed by the demand for turpentine and naval stores.

Back of this section, in the rolling Piedmont country, in the district around Hillsboro, German, English, and Scotch-Irish were settled, and behind them were the settlements already mentioned. The land was hilly, and, except upon the streams, not rich, though susceptible of high development through scientific agriculture. Intensive culture was demanded and not a great plantation system.

The extreme West, the mountainous region, was settled slowly, partly by the shiftless and incapable whom the economic pressure of population forces toward the free land, partly by the bold and adventurous spirits of whom Daniel Boone was a type. Such men, intoxicated by the sense of freedom growing out of their mastery of the forest, find the proximity of any neighbors unpleasant and go on to seek new lands. The difficulties of communication have kept that section more or less primitive to the present day, and it need not be considered as an industrial factor.

In 1790 these middle and western counties were almost self-sufficient. Land was plenty and cheap. Food was abundant, though from lack of markets there was little encouragement to raise more than could be consumed locally. The streams were not navigable, and the rough, hilly roads made transportation by wagon difficult and expensive. Each year wagon trains went to Philadelphia or Charleston, and later to Fayetteville and Cheraw; but only wares of small weight and considerable value could be hauled.

The domestic industries, which will be discussed more fully in another place, flourished. Though there were no towns of any size, the number and the skill of the artisans was such that, in 1800, it seemed probable that the logical development would be into a frugal manufacturing community, rather than into an agricultural state.

By the Constitution of 1776, the inhabitants of the Eastern counties (many of which were originally only precincts of the first counties) had a disproportionate share in the government. To the demands of the West for improvement in transportation and educational facilities, they turned a deaf ear. With them both local intercourse and communication with the world was easy, and they sent their children either to England to be educated, or to the Northern colleges already established.

Naturally, bitter jealousy and antagonism,

which are not yet entirely gone, grew up between the East and the West. The inhabitants were different in descent, in religion, and in habits of thought. The physical differences in territory fostered differences in social and economic organization which emphasized the distinctions already existing.

The Western section began to strive for the establishment of new counties, hoping thus to gain its ends. Sometimes the creation of new counties seemed to become an end in itself rather than a means. The whole legislative history of the state for fifty years is largely comprised in this struggle between the East and the West. All great questions were pushed aside for the engrossing dispute. Meanwhile economic and social interests, which might have been promoted by wise legislation, languished.

Except in the East, the feeling against slavery was strong during the first quarter of the nineteenth century. The Manumission Society was founded in 1816 and the name changed to the Manumission and Colonization Society the next year. Many abolition societies were also organized. At the meeting of the Manumission Society in 1825, 36 branches were reported, and in 1826 there were 1600 active members, many of them slaveholders. The Nat Turner slave insurrection of 1831, the growth of abolition societies in the North, and economic changes making slavery more profitable caused the dissolution of the society, and no meetings were held after 1834.[1]

The Western counties were also greatly affected by the increasing importance of cotton, and the number of slaves grew rapidly. On the Southern border where cotton was a profitable crop, and also in the rich river valleys farther north, a plantation system developed. A study of population figures indicates clearly these changes.

In 1790 the population of the West was: whites 136,655; slaves 30,068; while in the

[1] Weeks, *op. cit.*

East there were 151,549 whites and 70,508 slaves. In 1860 the white population of the West was 385,724 and the number of slaves 146,463; while in the East there were 244,218 whites and 184,596 slaves.

Thus the white population of the East had increased 61 per cent. in seventy years while its slave population had increased 162 per cent. The white population of the West had increased 182 per cent. and its slave population 387 per cent. during the same period. Obviously, the West was the growing section both in whites and slaves. Further, though the actual number of slaves in the West never equaled the number in the East, the rate of increase was much greater. The proportion of slaves to whites in the West never reached the proportion in the East, however. In 1860 the slave population of the West was 38 per cent. of the white, while in the East it was more than 75 per cent. Counting five persons to the family, it appears that there were for every white family in the

West 1.9 slaves, while in the East there were 3.83 slaves.[1]

One consequence of the extension of slavery was the emigration of thousands of the small farmers. Tennessee was settled from North Carolina. With the development of the Northwest Territory, that section received large additions, chiefly of those to whom slavery was obnoxious. The New Garden Monthly Meeting (Quaker) between 1801 and 1866 issued 245 certificates to individuals and families going to Ohio and Indiana. In the latter state one finds names of streams and townships directly transferred from North Carolina. It is estimated by Quaker historians that in 1850 one third of the population of Indiana was composed of North Carolinians and their children.[2]

There was also a strong current toward the Southwest. Georgia, Alabama, Mississippi,

[1] Consult Bassett, "History of Slavery in North Carolina," *J. H. U. Studies*, 1899.
[2] Weeks, *op. cit.* See also Marryat, "Diary in America" (1839), p. 143.

and Louisiana gained many settlers. Younger sons of slaveholders, taking a few slaves with them, or non-slave holders, made their way into that region where fertile lands might be had at nominal prices, and developed larger plantations than they had left. In 1855, 28 per cent. of the students of the University of North Carolina were from other Southern states, and in 1859, 39 per cent. of the students were from without the state, chiefly the sons of expatriated North Carolinians.

The emigration was at its height between 1830 and 1840. During that decade the white population of the state increased only 2.54 per cent. compared with 12.79 per cent. for the preceding period. There was no countercurrent of immigration to replace the loss. There has been no considerable addition of foreign population since the Revolution. In 1900 the proportion of the population born abroad was less than one half of one per cent., the smallest in the Union. As a result of this drain, the relative rank of the state in population declined from third in 1790 to twelfth in 1860. In 1900, 329,625 natives of the state were living in other states, while only 83,373 natives of other states had come in to take the place of the emigrants.

The migration to other states left large tracts of vacant land, and the state became more distinctly agricultural. The old manufacturing was incidental to agriculture, and the opening of railroad communication in the West after 1850 found other states ready to supply manufactured articles more cheaply than the local workmen could do. Though the cotton and tobacco manufacture slowly increased, the home industry was, as a whole, distinctly less successful.

The agriculture viewed by present standards seems wasteful. Since land was so abundant and so cheap, the usual plan was to work it until exhausted and then "turn it out" to be restored by the slow process of nature, the growth and the decay of vegetation. One may find to-day in tracts grown up with "old field" pine and sassafras the traces of corn or cotton ridges from which the last crop was harvested fifty years ago. Agriculture became more and more a matter of a few staple crops, and these the same which were grown to greater advantage in the new lands of the Western states, or by great gangs of slaves on the rich fields of the Southwest.

Educationally, the state did something. The Constitution of 1776 provided that "all useful learning shall be duly encouraged in one or more universities." The University of North Carolina, the second of the state universities, chartered in 1789, has a long and honorable history. Later the leading religious denominations each established a college. But the idea that universal education was one of the functions of the state was slow to develop, and the percentage of illiteracy grew to be the highest of the states.

A small fund for public education was created in 1825. A large part of the surplus distributed by Congress in 1836 was turned into this fund, and in 1840 a state system of public schools was instituted. In this year 14,937 children attended, but the number grew by 1850 to 104,095, a number equal to five ninths of the white population between six and twenty-one years of age. The schools grew in popularity and efficiency, and two more decades of uninterrupted existence would have shown a great impression upon the mass of illiteracy.[1] Since the wreck of the Civil War the educational progress has been marked. The public schools receive each year a larger proportion of the taxes. The colleges have grown in students and in equipment. Normal and technical schools for both races have been established. Institutions for the defective and unfortunate have been improved, but the burden of illiteracy is still tremendous.

Upon the Puritanism of the Scotch-Irish

[1] U. S. Census, 1850–1860, and Ingle, "Southern Side Lights" (1896).

has been grafted religious emotionalism. After the Revolution, French atheistic writers made a certain sort of materialistic philosophy fashionable; but the great revival of religion at the beginning of the nineteenth century swept the country. Camp meetings lasting for weeks and the discussion of abstract theological questions went on together. Communities and families have been rent by the excitement growing out of ecclesiastical disputes and debates.

Conflicting influences have hindered the progress of the state. Sectional jealousies have prevented concerted action, and yet there has been surprising unanimity upon great questions. The state has been conservative and slow, yet has led, rashly sometimes, in many things; it has been prosaic, yet capable of exhibitions of sentiment and enthusiasm beyond the ordinary.

The whole history has been a series of paradoxes. Restless under government imposed from without, it was quiet when the laws were suspended. So penurious that every Royal governor complained bitterly of his difficulties, the province built a palace for the governor which was pronounced the handsomest building upon the Western hemisphere. It was the first province to declare itself independent of Great Britain, and yet the twelfth to enter the Union.

When the whole expenses of the state government were $96,000 a year, the legislature appropriated $30,000 to buy a statue of Washington by Canova. In 1848, under the influence of Miss Dorothea Dix, a sum larger than the whole yearly income of the state was appropriated to build an asylum for the insane. The policy of internal improvements was unpopular, yet the state subscribed two of the three million dollars required to build the North Carolina Railroad.

A majority of the voters in 1860 opposed secession, and the question even of holding a convention was defeated as late as February 28, 1861. Not until forced to choose between fighting the North or the South was the ordinance of secession passed, May 20, 1861. When once engaged, the state furnished one fifth of the soldiers in the Confederate armies and strained all her energies to carry the struggle to a successful conclusion.

This is the state and these are the people who are now living in decades, whole centuries, of economic development.

## CHAPTER III

DOMESTIC MANUFACTURES AND THE BEGIN-
NING OF THE TEXTILE INDUSTRY

I

THE idea so industriously fostered that the settlers in the South were destitute of mechanical ability is entirely erroneous. The Scotch-Irish and the German immigrants brought their trades with them, and among the Moravians were artisans of every sort. The Huguenots and the Swiss included many skilled workmen. There was need. European goods were expensive and difficult to procure. The settlers had few products of their own sufficiently valuable to pay the cost of transportation even to the seacoast, to say nothing of the trip across the Atlantic. What could not be made locally must be foregone.

Take, for example, a house back from the seacoast. It was built of logs during the early years, with the floor of other logs split in half or even of clay. The open spaces between the logs in the walls were filled by poles and clay. The chimneys were of stone at the bottom, with flues of small poles daubed inside with clay. All hinges and fastenings were of wood. The iron pots in the fireplace and the coarse dishes upon the table were brought from Pennsylvania, but the rude furniture was made upon the spot. Later, in some neighborhoods, brick or stone houses supplanted the logs before the introduction of sawmills. Boards were hewn or sawn by hand from the trees. Mecklenburg County was prosperous, and the citizens were intelligent; but the first steam sawmill was not established until after 1850.[1]

When the first hardships of pioneer days were overcome and wants multiplied, a great variety of small industries sprang up in every neighborhood. Spinning wheels, made by local workmen, spun wool, cotton, and flax, which

[1] Alexander, "History of Mecklenburg County" (1902).

## DOMESTIC MANUFACTURES

looms, also made in the neighborhood, converted into cloth. Various goods were made from these three materials singly or in combination. Dyes from the fields and woods added a pleasing variety. Bedspreads and rag carpets of wonderful design were woven.

Speaking from personal knowledge of the twenty years before the Civil War, an old man says: —

"Almost every family had their own loom, wheel, and cards for every two female members of the family, white and black. Sewing thread was also spun, doubled, and twisted upon the spinning wheel at home. Only for very fine goods was spool thread bought. . . .

"Negro women spent all their time when not employed in making or gathering the crops, in spinning and weaving cloth to make their clothes or bedding, or clothes for members of the white family.

"A generation or two ago women took a delight in showing each other their fine handiwork. They knit most beautiful hoods and shawls. . . . All the clothing was made at home except wedding outfits, or for extra occasions. All the footwear was home made."[1]

Hats were made from fur, wool, or braided

[1] Alexander, "History of Mecklenburg County" (1902).

straw. The most influential man in the state in 1831, Nathaniel Macon, who had been president of the United States Senate, wore at a public gathering a suit of homespun, and a hat which his overseer's wife had made for him.[1] Of course, it was perhaps uncommon for a man of Mr. Macon's prominence to dress in homespun. The wealthier planters, particularly in the East, bought expensive broadcloth, silks, etc.; but even to-day, in the more remote sections, homespun is yet worn, rag carpets are woven, and the old women have not yet lost their skill at the loom.

Hides were tanned; boots, shoes, harnesses, were made by the farmer himself or by local workmen in exchange for meal or meat. In 1810, in the number of hides tanned, the state ranked fourth, and the art has never been lost. Even to-day many farmers mend the shoes for their families and make a part of the harness for the work animals upon the farms.

[1] Creecy, "Grandfather's Tales of North Carolina History" (1901).

## DOMESTIC MANUFACTURES 41

Some degree of skill in woodworking was attained very early. Furniture was made of oak, ash, cherry, black walnut, as well as the omnipresent pine. Chairs with seats of withes, rushes, or leather are yet made, though in recent years the cheapness of the factory product has practically superseded hand work. Baskets of every kind were made, and even yet the hampers used in cotton picking are the handiwork of some decrepit or crippled negro who preserves the old craft. Wagon makers built the heavy road wagons and lighter carriages, the latter not without some degree of elegance.

"Mr. H—— made vehicles upon honor. If he sold a buggy and harness, he would warrant it to stand three years; but he would charge from $150 to $200. His buggies were known to last, with ordinary care, from ten to fifteen years." [1]

Farming implements were made, including gins, gin presses, and the heavy wooden cog wheels for the transmission of power. On the old plantations the ruins of the gin machinery

[1] Alexander, "History of Mecklenburg County" (1902).

may yet be seen. Flails, fanning machines for cleaning grain, water wheels for saw or grist mills, were all made within the state. The cooperage establishments turned out buckets, barrels, vats, and firkins.

Bar iron from a local "bloomery" or perhaps "Sweet" (Swede) iron brought from Philadelphia furnished the material for nails, cut or forged by hand, horseshoes, plows, wagon tires, grain scythes, hinges, locks, etc. Near Lincolnton, after about 1822, was an ax factory, the product of which was widely sought on account of its excellence. In 1800 at High Shoals there were a rolling mill and shops which turned out various products from wrought iron, including bars, nails, plowshares, etc.[1] There were also establishments which made hollow ware, *i.e.* pots, kettles, etc. Tench Coxe in 1810 ranked the state second only to New Jersey in the number of "bloomeries."

Much machinery for the early cotton mills

[1] Tompkins, pamphlet (1902).

was made by the local blacksmiths. They were important men in the community and often grew prosperous. Some invested their savings in land and with the development of the state grew to be the holders of large and valuable estates. Their sons often went to college and became prominent planters or professional men.

By the streams where the clay was found to be particularly tenacious and smooth, pottery works were established. Little if any table ware was made, but crocks for the dairy, jars for household purposes, and jugs were made in abundance. The surplus beyond the neighborhood demand was peddled from wagons which visited the country stores or the individual buyers. Many of these little establishments endure to the present day, and the figure of the potter's wheel is intelligible to thousands who have seen the fashioning of the clay. The statistics for 1810 already mentioned [1] show many other indus-

[1] p. 11.

tries. There were eight manufactories of gunpowder, and two salt works. Six thousand pounds of paper were made in this year and rope walks were in operation. The distilling of ardent spirits was an important industry, and in the production of turpentine and varnish the state easily led. In the value of all manufactures, the state ranked seventh.

This rank in manufacturing was lost with the succeeding decades as agriculture assumed greater importance. In the East, where there was more wealth, and communication with the outside world was easier, reliance upon foreign goods became more pronounced; but until 1850 it is safe to say that a majority of the people in the Middle and Western counties dressed chiefly in clothes of domestic or local manufacture, lived in houses furnished by the local cabinetmaker, rode in vehicles made within the state, and used implements made in the neighborhood.

## II

The first cotton mill in North Carolina and one of the first south of the Potomac was built about 1813, on a small stream near Lincolnton, which is now a considerable manufacturing town. Lincoln County had been settled principally by Germans, Scotch-Irish, and Swiss, many of whom had mechanical ability. Michael Schenck, a native of Lancaster County, Pennsylvania, who had prospered in his new home, determined to build a mill. Some of the machinery was purchased in Providence, Rhode Island, and was hauled by wagon from Philadelphia. Other parts were made by Schenck's brother-in-law, a skilled worker in iron. The first dam did not hold and it was necessary to rebuild it lower down the creek. A contract with a local workman for the construction of additional machinery is in the possession of one of the Schencks' descendants. It reads as follows:[1]—

[1] Schenck, "Historical Sketch of the Schenck and Bivens Families" (1884).

"Articles of agreement, made and entered this 27th. day of April 1816, between Michael Schenck & Abraham Warlick of the County of Lincoln and State of North Carolina, of the one part, and Michael Beam, of the county and state aforesaid, of the other part witnesseth; that the said Michael Beam obliges himself to build for the said Schenck & Warlick, within twelve months of this date, a spinning machine with one hundred and forty-four fliers with three sets of flooted rollers, the back set to be of wood, the other two sets to be of iron; the machine to be made in two frames with two sets of wheels; one carding machine with two sets of cards to run two ropings each to be one foot wide, with a picking machine to be attached to it with as many saws as may be necessary to feed the carding machine; one rolling (*sic*) with four heads. All the above machinery to be complete in a workmanlike manner. And the said Beam is to board himself and find all the materials for the machine and set the machinery going on a branch on Ab. Warlick's land below where the old machine stood; The said Schenck and Warlick are to have the house for the machine and running gears made at their expense, but said Beam is to fix the whole machinery above described thereto; the wooden cans for the roping and spinning and the reel to be furnished by said Schenck and Warlick; all the straps and bands necessary for the machinery to be furnished by said Schenck and Warlick.

In consideration of which the said Schenck and

Warlick are to pay to said Beam the sum of thirteen hundred dollars as follows, to wit: three hundred dollars this day, two hundred dollars three months from this date, one hundred dollars six months from this date, and the balance of the thirteen hundred dollars to be paid to the said M. Beam within twelve months after said machine is started to spinning. In testimony whereof we have hereunto set our hands and seals, the day and year above written.

|  |  |  |
|---|---|---|
|  | ABSALOM WARLICK. | (SEAL) |
| Test, | MICHAEL SCHENCK. | (SEAL) |
| ROBT. H. BURTON. | MICHAEL BEAM. | (SEAL) |

The mill was prosperous, and John Hoke and James Bivings bought a share in 1819. The firm erected a larger mill of three thousand spindles, the Lincoln Cotton Factory, on the South Fork of the Catawba, about two and a half miles south of Lincolnton. Attached to this mill was an annex, which made various articles from iron. Wagons came from a distance of a hundred miles to secure yarn, and the mill continued in successful operation until burned by an incendiary in 1863. On the site the Confederate government erected a laboratory for the manufacture of medicines, and twenty years after the war a cotton mill again began operations.

In 1820 Colonel Joel Battle, the grandfather of Professor Kemp P. Battle of the University of North Carolina, opened the Rocky Mount Cotton Mills in Edgecombe County, with more than two thousand spindles. Coarse yarn for neighborhood consumption was spun here by negroes. Nearly all of them were slaves, belonging to the mill owners, or to their neighbors, though a few free negroes were employed. White labor was substituted in 1851.[1] The mill, though making only twelve to fifteen hundred pounds of coarse yarn, 4's to 12's, daily, was much hampered by lack of a steady market.

Apparently the first application of steam to the industry was at the Mount Hecla Mills at Greensboro about 1830. The machinery for this mill was shipped from Philadelphia to Wilmington, then up to Cape Fear River to Fayetteville, and was hauled across the coun-

[1] See Chapter XIII.

try by wagon. When wood for fuel grew scarce, the machinery was moved to Mountain Island, where it was run by water power.

Soon after 1830 E. M. Holt, one of the most successful manufacturers the state has known, built a mill on Alamance Creek. Finding difficulty in disposing of all his yarn, he began between 1850 and 1860 the manufacture of coarse, colored cloth known as "Alamance plaids." Success attended the venture and the product became more than locally known. To-day, throughout central North Carolina, "Alamance" is almost universally used as a synonym for the coarse ginghams on the shelves of the country merchants. Other mills were built by him and his sons, and the family is prominent in manufacturing at the present time.

In 1840 Francis Fries, a descendant of a Moravian minister, who had had some experience in cotton manufacturing as agent of the Salem Manufacturing Company, began a small wool business. To this was added dyeing vats to color the cloth woven by the farmers' wives, and later spindles and looms were added.

Other mills had been built during the decade, and in 1840 twenty-five establishments were reported to be in operation. The total number of spindles, however, was only 47,900, with 700 looms. The number of operatives was only 1200, the capital $995,300, and the consumption of cotton 7000 bales, totals surpassed by single establishments at the present day.

During the next twenty years the number of establishments increased, though the spindles decreased. In 1860, 39 mills with 41,900 spindles and 800 looms were reported. The consumption of cotton is given as 11,100 bales, the capital as $1,272,750, and the number of operatives as 1755. It is noteworthy, however, that only nine of these establishments were in the Eastern counties. The increase in cotton consumption is probably due to more regular operation. Many of the early mills ran only a part of the year. The water power was often imperfectly utilized, and the mill was necessarily stopped when the streams were abnormally high or low.

Often the mill was stopped when the neighborhood demand was satisfied. Commercial organization was lacking. Little attempt to secure more than a local market seems to have been made. Instead of selling the whole product to a distributing agent, each mill was its own distributer and depended chiefly upon local demand and upon accidental outside consumers. A third difficulty was the fact that the operatives were such only incidentally.

Upon Deep River in Randolph County, where five mills were built before 1850, conditions were somewhat peculiar in this respect. These mills were in a section where the Quaker influence was strong. Slavery was not widespread and was unpopular. The mills were built by stock companies composed of substantial citizens of the neighborhood. There was little or no prejudice against mill labor as such, and the farmers' daughters gladly came to work in the mills. They lived at home, walking the distance morning and evening, or else boarded with some relative or friend near by.

The mill managers were men of high character, who felt themselves to stand in a parental relation to the operatives and required the observance of decorous conduct. Many girls worked to buy trousseaux, others to help their families. They lost no caste by working in the mills. Twenty years ago throughout that section one might find the wives of substantial farmers or business men who had worked in the mills before the Civil War. Some married officials of the mills.[1]

In many localities, however, there was difficulty in securing the necessary labor, arising not so much from the feeling that such labor was degrading, as on account of the confinement and the necessary subordination. The people had been accustomed to out-of-door life for generations. Life was simple, and discontent with the loneliness of the farms had

[1] For somewhat similar conditions in New England in the thirties, see Robinson, " Loom and Spindle" (1898).

not assumed its present proportions. To work indoors seemed too great a sacrifice.

The spirit of independence was strong in the rural population. They felt themselves as "good as anybody," and disliked to take orders. They did upon their own farms labor of the same sort, and much that was more unpleasant; but this was done for themselves. Both men and women worked for wages for their more prosperous neighbors, but their position was not distinctly menial. They were not so much working *for* that neighbor, as they were working *with* him, assisting him and his family.

Such workers were not considered servants, but ate at the family table, and occupied rooms in the house. Working in a mill under overseers seemed to many a sacrifice of independence, and any curtailment of personal liberty was resented. On one occasion, the attempt to prevent operatives from looking out of the windows, by painting the glass, would have resulted in a general strike but for the restoration of the clear glass.

Further, the large emigration had left many vacant farms, and there was abundant room for all upon the soil. As the state came more and more under the influence of the plantation system the ambition of every farmer, however small, was to become a planter. To go to the mill with the intention of remaining meant the definite abandonment of such ambition, and few were willing to make that sacrifice.

## CHAPTER IV

THE GROWTH OF THE INDUSTRY SINCE 1861

THE beginning of the Civil War found the state with less than $1,500,000 invested in cotton manufacturing, possibly $300,000 in wool, and as much in iron. Yearly she was growing more dependent upon the North and upon Europe, not so much from the decay of the industries already existing as from lack of their expansion. The home manufactures had not kept pace with the increasing wants. In some industries they had actually declined. "Yankee Notions," in increasing quantities, were imported following the increasing reliance upon cotton growing.

Within twelve months after the beginning of the war the state became as it had been in 1810, to a great extent, self-sufficient. The cotton, woolen, and leather manufactories were taxed to their utmost capacity. Spinning wheels and looms which had been retired to the attics were again brought into service. With the purchase by the state, under direction of the great war governor, Z. B. Vance, of the steamer *Ad-Vance*, state direction was added to the activity. This boat, which made eleven successful trips through the blockade before it was captured, brought in many things which were sorely needed. Sixty thousand pairs of hand cards for preparing cotton and wool, machinery for manufacturing shoes, textile repairs and supplies, were included in the cargoes.[1]

An account of the efforts and expedients of the people during that period would make a book of intense interest. Nothing was wasted. The law forbade the distillation of grain into alcoholic beverages. Luxuries were foregone, and for every supposed necessity no longer procurable a substitute was found.

[1] For an interesting account of state activity, see Governor Vance's article in "History of North Carolina Regiments," Vol. V (1901), also printed in Dowd, "Life of Vance" (1897).

## GROWTH OF THE INDUSTRY SINCE 1861

The cotton manufacturers, in striking contrast to their course elsewhere, did not take advantage of the increased demand to pile up fortunes for themselves. Almost without exception they refused to sell their product to speculators. They were usually men of standing and influence in their neighborhoods, and valued the respect in which they were held. The value of public opinion as an economic force in the South has never been properly estimated. It is a power to-day, and the man who demands his pound of flesh from a helpless, unfortunate neighbor is censured for his harshness, rather than praised for his exactness. Social aversion makes his position like that of the usurer. Some primitive ideas of the duties toward neighbors still prevail.

The course of General W. H. Neal of Mecklenburg County is perhaps typical. He owned a little mill, containing only 500 spindles and a few looms, which had begun operation in 1850. When the demand for yarn exceeded the supply, he adopted the plan of considering the absolute necessities of applicants, rather than their desires. The soldiers' widows came first, and even his own children were forced to take their chances with other applicants. He was paid in Confederate money, though speculators offered to pay in gold. The fate of this mill was that of several others. In 1866 the machinery was so worn that further operation was unprofitable, and a grist mill took its place.

When the state was overrun with Federal troops, a number of mills were destroyed. Among them were the Rocky Mount Mill, burned in 1863, and the plant of the Richmond Manufacturing Company, which was burned by Sherman's army in 1865. This mill, which had been in operation since 1833, was rebuilt and enlarged in 1869, and has been in successful operation ever since. Five mills, in and around Fayetteville, were also burned in 1865, by order of General Sherman.[1] Stoneman's raiders also burned the mill at Patterson, Caldwell County.

[1] Vance, "Last Days of War in North Carolina," in Dowd, "Life of Vance" (1897).

The mills which escaped destruction before peace was declared were generally in poor condition. The machinery, tried by the strain of years, was worn and much was obsolete. Some owners were ruined by emancipation and the disarrangement of the whole economic system, and had no capital for renewal. The country was prostrate, the future was uncertain, and the outlook was dark. Some mills, sold at auction, brought sums so small that profitable reorganization was possible. Others could find no purchasers and were stopped entirely. Generally, however, the mills continued to run.

The high price of cotton, and the development of the tobacco industry, in the years immediately following the war, brought some money into the state which was almost without a medium of exchange. Indeed, the abnormal price of cotton as a factor in the recovery of the South has not been sufficiently emphasized, even though many evils followed in its train. Though cotton had been growing steadily more important, it had not been the sole crop. On every farm and plantation, grain and meat had been produced. Now the whole energy was turned into cotton growing wherever it was possible. The other factors, which also influenced this change, will be discussed in another place.

In 1870 we find enumerated only 33 mills with 39,900 spindles and 600 looms. The capital invested was $1,030,900, and the consumption of cotton had dropped to 8500 bales. The average number of spindles to the establishment had risen to 1210 — an advance of nearly 20 per cent. which would seem to indicate that the smaller and less economical establishments had not survived.

During the next decade hope began to return. The reconstruction government, while corrupt, was less greedy than in other states. Some of those profiting by contracts and bond issues invested their gains in industrial enterprises. The great panic of 1873 did not, at first, affect the state severely. The state was so largely agricultural, so little money was invested in manufacturing, and there were so few banks that the first shock was not severe. Later the general depression was felt in the price of cotton. The high prices had caused the emphasis to be laid upon large, rather than upon economical, production. The price declined almost steadily from 23.98 cents the pound in 1870 to 10.38 cents in 1879, due partly to increased production but also to decreased demand.[1] Farming was less profitable, particularly as crops were short for several years; money was scarce, but still there was less suffering than was experienced in other sections.

During this decade the textile industry increased. Mills no longer distributed their own product, and much yarn was shipped from the state to be woven elsewhere, though the number of looms was tripled. Forty-nine establishments with 92,400 spindles and 1800 looms were reported in 1880, and the average number of spindles to the establishment reached 1890, an increase of more than 50 per cent. The

[1] Hammond, "The Cotton Industry" (1897).

capital reported was $2,855,800, and the consumption of cotton is given as 23,700 bales.

The best farmers had made money growing cotton, and began to invest some of the proceeds in cotton manufacturing. New mills were built, all of which did not succeed. Some managers yielded to offers, tempting on their face, to install machinery which had been used in New England. Farsighted manufacturers there, seeing the possibility of competition in coarser numbers of yarn, began to turn their attention to finer yarns, or else wished to take advantage of new inventions. Machinery, some of it, at least, in good condition, was offered at very low prices to the Southern mills. The result was generally not satisfactory, and as a result a few mills went into bankruptcy.

Much gratuitous advice was now offered to the section. It was gravely announced by those interested in preventing manufacturing development that the Southern climate was not suitable for spinning on account of the dryness; that machinery could not be kept in good condition on account of the moisture; that labor could not be found at all; that the native labor could never attain satisfactory skill; that it would be an economic waste to draw labor from the production of cotton into its manufacture. It was prophesied that the necessary managing ability could not be found, and capital was warned not to trust itself in the hazardous enterprise. Extreme solicitude for the savings of the South was also manifested.

But the habit of mind of the Southern people was changing. Those who had saved land and capital from the wreck of the war, or had gained them since, began to tire of the never ceasing contest with the inefficiency and unreliability of the freedman. As the older negroes who had been trained under the discipline of slavery became superannuated, it was found difficult to secure efficient laborers. The younger negroes preferred to work in gangs in the turpentine forests, at railway construction, or not to work at all. Hundreds of well-to-do farmers, disgusted with the struggle, practically abandoned their farms and moved to town, there to seek profitable occupation and investments. The country merchant also began to dream of managing greater enterprises.

Sometimes they were surprised at their success. Commercial and industrial ability was found not so rare as had been supposed. As their interests grew, ability to manage them was developed. The old idea of comfort — life upon a plantation — was no longer unchallenged. But more than this, the people generally began to be convinced of the probability of Southern industrial success. The awe of the ingenuity of the thrifty Yankee was no longer so pronounced. The people began to be willing to invest their surplus or savings in something other than a land mortgage.

Under a plan which will be described elsewhere, these savings, individually small, but large in the aggregate, were poured into cotton manufacturing. Towns in which not a single man could be accounted rich even by the modest standard prevailing, began to discuss the erection of manufacturing establishments. The process was not rapid. Inertia, timidity, inexperience, were to be overcome; but after 1890 the building of mills went on with increasing rapidity. The majority were small neighborhood affairs, but they were profitable.

The manufacturers of New England generally did not realize the revolution that was taking place in Southern life. That section had crippled or destroyed industrial enterprises existing in the South before the war, and it was difficult to believe that there was any menace to her supremacy in the textile industry. But the new Southern mills were not the same old wasteful establishments. New plants were built from the profits of the old. The newest machinery was installed. The unprofitableness of second-hand machinery was recognized, and only the best was bought.

The machinery houses began to take great interest in the development. Agents were sent to encourage building, and favorable terms of payment were granted. When sufficient capital seemed difficult to secure, the manufacturers of machinery offered to take a part of the price of the machinery in stock.

The statistics for 1890 show plainly the progress of the industry. The number of establishments was 91, and the number of spindles, 337,800, was more than three and a half times the total of ten years before. The number of looms, 7300, was more than four times as great. The capital reported as invested was $10,775,100, and the consumption of cotton, 107,100 bales, was nearly a third of the state's production.

With the publication of such statistics the attention of the North was fully aroused. The statement that the South could not manufacture cotton successfully was no longer heard. The march of events had proved the falsity of that prophecy. Some Northern manufacturers began to erect branch mills in the South. Few of these, however, came to North Carolina, but were located farther south, nearer the center of the cotton belt. The state had learned to rely chiefly upon its own endeavors, and ceased calling for outside capital to develop its resources.

Not every new mill was advantageously located. Nearly every little town in the central or west-central portion of the state built a mill or mills. Some were built away from railroads, sometimes to utilize water power, sometimes to secure cheap fuel or abundant labor. These advantages were often apparent rather than real, or at least temporary, or unreliable.

Profits, however, seemed almost certain. Mills, though not always economically managed, paid good dividends, and the best were phenomenally successful, though sometimes at the expense of a reserve for depreciation. The whole profit, in many cases, was paid to the stockholders and nothing was retained to replace worn machinery, nor to provide a reservoir from which dividends might be paid in a less profitable season.

The result, following these large dividends,

was almost a craze for mill building. Mills were too often built to produce those particular yarns which were most profitable for the moment, without due consideration of permanency of profit; or the interested advice of some commission house which made a specialty of certain numbers was followed; but as the managers have learned more of the business, the tendency toward finer numbers has been well marked.

The statistics given for 1900 illustrate the tendency. In 1890 the state produced 41,972,080 pounds of yarn below number 20, and only 3,076,558 pounds between number 20 and number 40. None finer than number 40 was reported, and the amount above number 20 was only 7.09 per cent. of the total production of the state. In 1900 the proportion was much changed. The number of pounds under number 20 was 99,021,341, but the amount between number 20 and number 40 was 56,527,998, while 886,200 pounds of yarn finer than number 40 were also reported.[1]

[1] U.S. Census Bulletin 215.

Before 1890 the question of satisfactory labor had not been entirely solved. The better class of population was not easily drawn from the farms to the factories. After 1890 the price of cotton, owing to increased production both of the domestic staple and of Egyptian and Indian, and also to the depression following the panic of 1893, went lower and lower. On the bulk of the crops of 1894 and 1895 the farmer realized little more than five cents, while much was sold below this low price, which was less than the average cost of production. Low prices for tobacco, corn, and wheat accompanied the ruinous price of cotton.

These unprecedentedly low prices of their products brought much distress to the farming population. Crops brought hardly more than fertilizer bills, allowing nothing for labor. Live stock brought less than the cost of feeding, even at the prevailing low prices of hay and grain. To secure the cash to pay taxes was a difficult problem. Debts incurred when times were easier were now a crushing burden.

A mortgage, once easily carried, was now an impossible load. Farms were sacrificed for a small part of their supposed value. The political revolution growing out of the prevailing discontent will be discussed in another place.[1]

Meanwhile the cotton mills seemed the only enterprises unaffected by the prevailing depression. The mills were running at their full capacity, often both night and day; were selling their yarns at a profit in Philadelphia, New York, and Boston, and were sending cloth to the Orient, and a limited quantity to South America. To the mill towns turned the discouraged from the farms, hoping for better times in industry than in agriculture. Renters and laborers went to those places where there was work with money wages for all. Landowners also sought employment. In some neighborhoods the movement assumed almost the proportions of an exodus. Among the migrants were the lazy, the shiftless, and the incapable, but there was also the hard-working,

[1] Ch. X.

honest element, which hoped to better its condition by industry.

As new mills were completed they were as quickly filled, even though situated out of the cotton country. When expenses of moving are offered in addition to wages, operatives can be drawn from other mills, though careful selection cannot always be exercised. The mobile labor is usually the unsatisfactory labor. Much of this willingness to move is due, however, to lack of adjustment to surroundings of the population, so lately taken from the soil. The new life cramps them at some points, and they move in the vain search for the freedom of the old, together with the advantage of the new.

With the return of higher prices for agricultural products, together with certain agencies tending to make life in the country more attractive, the movement toward the mills has become slower. Greater inducements are necessary to attract the new material from which efficient workers may be made. During the summer of 1904 many mills were without their full complement of operatives. Wages in North Carolina mills have been seldom cut. Nearly every advance has been permanent. So, in order to avoid formally raising the rate, which might be difficult to reduce, should the scarcity prove only temporary, some managers adopted an ingenious substitute. This was to pay the operatives for running more machines than could be efficiently operated. For example, a spinner capable of managing four "sides" would be paid for six, though two were only nominally in operation.

Meanwhile the building of new mills had gone on rapidly, and the average number of spindles also was increased. Mills to spin finer yarn and "specialties" were built, and finer cloth, both white and colored, was produced. Here again the prophecies from New England, that the Southern mills must confine themselves to the coarser grades, were disproved.

Though the trouble in China reduced the profits by curtailing the markets of the existing mills, the promoters were not frightened. The movement reached its climax about 1903, when twenty-nine mills were in process of construction. Since their completion there has been a cessation of building activity. The market has been glutted at times, partly owing to the troubles in the East already mentioned; partly to overproduction of certain numbers and grades; partly to the decreased demand resulting from higher prices of cotton. Mills with established reputations, however, have continued to make profits, although others have become bankrupt.

## CHAPTER V

#### THE PRESENT STATE OF THE INDUSTRY

THE traveler through some parts of North Carolina is seldom out of sight or hearing of a cotton mill. The tall chimneys rise beside the railway in nearly every town. Side tracks from the main line lead to the low brick mills and the clustering tenements, set down in fields where crops grow almost to the doors, or in the forest where a clearing has been made.

The state has more separate establishments than any other. Almost one fourth of the mills in the United States are within its borders, though in production it is only third. There are no great establishments like those in Massachusetts, New Hampshire, or even South Carolina, which count their spindles by the hundred thousand. The largest of the 263 cotton or woolen mills reported in 1904 has only 75,000 spindles and about 2000 looms; and

## THE PRESENT STATE OF THE INDUSTRY 75

this is really two mills more than a mile apart, though under the control of the same corporation, and the second and larger was built from the profits of the first. Some mills are so small that successful operation would seem impossible. One woolen mill has but 168 spindles and no looms, while one cotton mill has only 820 spindles. The average number of spindles is a little over 8000,[1] and, if the looms were divided, the number to each mill would be 185. This number could not possibly consume the product of these spindles during daylight, to say nothing of the large production at night.

Many of these mills, however, have no looms at all, but sell the yarn produced. Around Philadelphia especially, and to a less extent in New England, are many mills which weave only and buy their yarn. In such mills the variety

[1] 2,178,964 spindles ÷ 263 = 8285.
48,612 looms ÷ 263 = 185.
The discrepancy between the number of establishments reported by the Census and by the State Bureau is explained by the fact that the former reports mills owned by the same corporations as one establishment.

of goods woven and the frequent changes make the production of all the numbers and qualities unprofitable or impossible. Some of the North Carolina mills sell their product to others in the neighborhood. Where one family or one interest controls several mills, one establishment has often been built to consume the product of another.

These mills are located in fifty-four of the ninety-seven counties of the state, in every section from the seashore to the mountains. Much the largest number, however, is in the central or west-central sections. Here the mills are thickest. Gaston County, with a population of 27,903 in 1900, has 32 mills; Alamance, with a population of 25,565, has 23; Guilford, with 39,074 population, has 10 mills, some of them very large; and Mecklenburg, with 55,268 population, has 19. Only 34 are to be found in what are classed as Eastern counties, and only four in the extreme West.

The older mills were usually located upon streams to utilize the water power, but a

## THE PRESENT STATE OF THE INDUSTRY 77

drouth was so annoying that the majority of these have installed steam plants for auxiliary use at least. Comparatively few mills are in the larger towns unless the towns have grown up around them. Generally they are built upon the outskirts of a village. Considerable land is needed for buildings and tenements, and this is secured at farm prices. The operatives are thus separated from whatever distractions the town may afford, and the payment of town taxes is avoided. Many mills are in the country, though generally near a railroad.

All the buildings are well constructed of brick or stone, and the newer ones are seldom more than two stories high. Light is admitted from three or four sides, and often from the roof as well; and the circulation of air is free, in striking contrast to some old New England mills. There, in the summer of 1903, I visited a mill where electric lights were burning at noon, though the sun was shining brightly outside. The air in the mills which spin the finer yarns is kept moist by humidifiers, which throw out water in a fine spray. In summer the spinning rooms are pleasanter than the offices or stores near by.[1] In winter, the contrast between this humid atmosphere and the cold winds outside is severe. In the mills without humidifiers the temperature under the tin roof may reach 100° on an August afternoon. The construction of the buildings and the installation of automatic sprinklers reduce the risk from fire to a minimum.

The equipment of the newer mills is the best. Fewer mills here run obsolete patterns in machinery than in New England, and all instruments of production are better, because newer. Every improvement, every labor-saving device, is installed. The time is long gone when Southern mills are equipped from the scrap heaps of other sections. The expensive Draper-Northrop loom which saves one half to two thirds of the labor in weaving plain goods is extensively used, while its introduction

[1] For confirmation, see Young, "American Cotton Industry" (1903), p. 67.

## THE PRESENT STATE OF THE INDUSTRY 79

into New England has been proportionately much slower. There managers of mills have felt that they could not afford to scrap their ordinary looms, perhaps running better than new, to invest in this expensive invention, which is not yet entirely perfected. Meanwhile the Southern mills which have installed them are reducing materially the labor cost, and with it the profits of the New England mills.[1]

The product of the North Carolina mills is yarn, "gray" (unbleached) cloth, plaids, ginghams, denims, toweling, canton flannel, hosiery, etc. By far the greater number of them are employed in the production of coarse cloth and the coarser numbers of the yarn, from 12 to 24.[2] From some mill or other, however, almost every standard product of cotton may be procured. The coarser yarns require less

---

[1] The Boott Mills at Lowell have just been reorganized (1905) after failure largely due to neglect to keep abreast of recent improvements in machinery, if the current reports are to be trusted.

[2] For classification of yarns, see Ch. VIII, p. 132.

skill in the manufacture, but with the increasing competition in these grades the tendency toward the finer numbers is steady. The Avon Mills in Gastonia spin number 60's from Egyptian cotton, and the Daniel Mill at Lincolnton has spun from combed sea-island cotton the finer numbers up to number 100.

There has been little difficulty in securing labor capable of the manipulation of fine goods. Of course, operatives fresh from the farms cannot at once display the requisite dexterity. By selecting those already trained upon coarser goods, as individuals, rather than employing whole families, success has followed. The mills making fine goods are necessarily confined to mill centers, where a large body of operatives is present from which selection may be made. Thus two predictions of Edward Atkinson have been disproved: the one that Southern cotton mills could not be successful; the second and later that only coarse goods could be made.[1]

[1] Address printed in "Report of Director General International Cotton Exposition," at Atlanta (1882).

The development of the cotton industry in North Carolina is a striking instance of the manner by which a people in poor or moderate circumstances can establish manufactures. Little foreign capital has been invested in North Carolina, contrary to the condition in South Carolina, Georgia, and Alabama. It is manifestly impossible to secure the residence and holding of every stockholder; but those best informed estimate that 90 per cent. of the capital has been invested by residents of the state. Further, the Northern capital has come chiefly since the success of the mills has been assured. The industry is distinctly a home enterprise, founded and fostered by natives of the state. During the ten years just past, several large mills have been built with foreign capital, but they have not greatly changed the proportion. A larger amount of such capital has been invested in mills already in operation, or has enabled a successful manager to enlarge his plant.

The ownership of the mills is widely distributed. While there are many in which a single man, or a single family or group, owns the whole, or a controlling interest, as, for example, the Holt family in Alamance and Davidson counties, which owns more than a dozen mills, the stock in the majority is widely distributed, owing to the method of building, which has often been an installment plan, on the following order: —

The subscription to the shares (usually of a par value of $100) is made payable in weekly installments either of 50 cents or $1 the share, without interest. Occasionally a mill has been built with a 25 cent installment. Experience has shown, however, that this requires too long a period, as nearly eight years is required to pay the stock in full as against four or two years for the larger sums. Those having ready money may pay the whole amount at once less 6 per cent. discount for the average time, making the stock cost $89.60+ in cash. Usually nearly or quite a year is required to construct the buildings. The installments more than suffice to pay the expenses, as real estate and buildings rarely cost more than 20 per cent. of the capital stock. The installments and the amount paid by those who have taken advantage of the discount is placed in some bank, which is thus put under obligations to the mill, and besides has a lively anticipation of business to come. Often the directors of the mill are also stockholders or directors of the bank. Machinery may be bought on long credit, six, twelve, or even eighteen months, with interest at 6 per cent. after delivery. Sometimes the makers of machinery have taken a part of the cost of the machinery in stock, and in a few instances the commission houses have also subscribed in order to control the product. There has seldom been any bonded indebtedness intended to be permanent.

Profits in the past have been so large that often before the last payment on the stock is due, a sum sufficient to pay all obligations has been accumulated. One especially successful mill of this class, organized with a capital of

$100,000, secured the buildings of an unsuccessful woodworking establishment, which with alterations and additions were adequate for the purpose. The installment was fifty cents the week on each share. When $35 a share had been paid in, in seventy weeks, a dividend of 4 per cent. on the capitalization was declared, and it has never failed to pay either 4 or 5 per cent. each half year since. Further, a large addition has been built and set in motion from the profits of less than ten years' operation. This is by no means universal. Some have not paid dividends for months after the stock was entirely paid in, and a few have never been profitable.

At first, the stock is widely distributed. Bankers, merchants, physicians, clerks, lawyers, teachers, mechanics, and even operatives in other mills subscribe. When difficulty is experienced in securing the desired amount, subscriptions of one share may be accepted. The average holding is seldom above $1000. This widest distribution does not last, of course.

Some subscribers find difficulty in keeping up their installments and transfer their subscription to others; some grow tired of waiting for dividends, which seem slow in coming. Some, who have used their subscriptions as a savings bank, sell in order to buy a home, or to start in business for themselves.[1] The stock tends to become concentrated in fewer hands, though a small body of men seldom secures control of a successful mill of this class. After a time a contrary centrifugal tendency develops through division of estates, business changes, etc., as the stock is almost invariably held for investment and not for speculation. If a mill is unprofitable for several years, a few men may gather in the stock on the chance of a successful reorganization. The North Carolina mills have been almost invariably managed honestly in the interest of all the stockholders. Seldom

[1] The frequency with which new mills have been started, and the success of the local Building and Loan Associations, have a decided effect upon the size of deposits in savings banks. On this point, see the testimony of S. Wittkowsky before the United States Industrial Commission, Vol. III.

have those directly in control attempted to "freeze out" the small investor, though such instances in other Southern states are not unknown.

The effect of this wide ownership influences public opinion in several directions. The attitude of any rural or semi-rural community toward the larger corporations is generally hostile. This state is no exception, as the verdicts in damage suits against railroads and telegraph companies plainly show. Toward the cotton mill, however, the attitude has been decidedly friendly. Boundary lines have often been changed to throw a proposed establishment outside the town or village limits, for a time at least. The mill thus escapes the payment of town taxes until it is well established, and often for a considerable period thereafter. Thus many communities have a considerable population which really belongs to the towns, though it does not appear upon their census or tax returns.

This attempt to lighten the burden of taxation is shown in other ways. Though the law demands that all property shall be assessed at its true value, it has been generally understood that the assessment of real estate, live stock, etc., is not more than two thirds to three fourths of the real value. The same principle has been applied to the mills. Formerly mills, which possibly had built large extensions from surplus, would be assessed simply upon the capital stock. A mill, the market or book value of which amounted to 100 per cent. advance on the capitalization, might pay taxes upon only three fourths of the capital stock. Though the method of assessment has been changed, the mills do not yet pay taxes upon their market value. This has not been done by the collusion of corrupt officials, but by common consent.

In many other ways the mills have been favored. The motive of many investors has been not only to secure a profitable investment but to "help the town." The pay roll of a $100,000 mill — a favorite size — ranges perhaps from $200 to $350 per week. The value of the cotton consumed weekly, depending of course upon the price and the fineness of yarn, is between $400 and $3000. Both the operative and the farmer spend a large proportion of this sum in the town. The money paid for fuel and other supplies is often large, and the influence of this expenditure in a small town is enormous. It is the general sentiment that such a stimulus to trade must be fostered. This attitude of friendliness is changing in some sections, however, and suits are more frequent.

The profits in the North Carolina mills have been large. There are, of course, as many rates as there are mills. The best authority upon cotton manufacturing in the South, Mr. D. A. Tompkins of Charlotte, estimates the average net profits for a period of twenty years up to 1900 at about 15 per cent. Since that year the average rate has probably been less. Some mills have made much more. Instances of 40 to 60 per cent. dividends are not unknown. In such mills, however, the plant has been enlarged from profits without proportionately increasing the capitalization. A mill which does not accumulate a surplus suffers during a less profitable period, however. Even when a surplus appears upon the books, it is often more apparent than real, since proper allowance is not always made for depreciation. The mills are so new, and so little is known of accounting, that the absolute necessity of providing a fund to replace equipment, when worn or obsolete, has not been realized in every case.

The unsuccessful mills are often so because of slavery to the commission houses through which they sell their product. Too many Southern mills have been built with insufficient working capital or with none at all. The commission houses charge 4 per cent. on unbleached cloth, and 5 per cent. on yarns and fancy cloths, and sell when and to whom they please. Goods are sold upon sixty days' time, with 2 per cent. discount for cash within ten days. The commission houses, many of which have banking connections, gladly advance

75 to 90 per cent. of the market value of unsold goods, charging the mill double the rate of interest which they themselves must pay for the money. Thus interest charges often eat up profits. The commission house to which the mill is indebted may demand entire control of its output, and the manufacturer may not receive in every case a price as high as might be realized in a market entirely free. Mills without adequate capital have succeeded only because there has been generally a large margin between cost of production and the average selling price. With the great increase of competition, the mill handicapped by debt from the beginning finds successful operation increasingly difficult.[1]

An apparent contradiction of economic law is found in the fact that the profits of the smaller yarn mills have seemed to be greater than those of the larger establishments. While, owing to the more careless accounting in the smaller

[1] See Tompkins, "Cotton Mill, Commercial Features," p. 128, and Young, "American Cotton Industry," p. 117.

mills, some of this excess does not really exist, all the difference cannot be thus explained. The small mills are usually in the country or in the small towns. They draw their cotton from the surrounding territory, and may save slightly in freights over the mills which must draw a part of their supply from other states. In coarse goods the largest cost is the raw material.[1] Purchasing cotton in small quantities is no disadvantage as the bale is the unit, and five may be purchased at the same price per pound as a hundred or a thousand. Many of these mills have burned wood from the surrounding country. One and a half to two cords of wood is estimated to produce as much steam

[1] The following table, calculated by Mr. D. A. Tompkins, shows the relation: —

PERCENTAGE OF THE TOTAL COST OF FINISHED PRODUCT ATTRIBUTABLE TO THE FACTORS OF MATERIAL AND LABOR

|  | Cotton | Labor |
|---|---|---|
| United States | 44% | 26% |
| New England | 42% | 28% |
| South | 59% | 19% |

A slight difference in the price of cotton may make a great difference in the rate of profit.

as a ton of coal. When wood is purchased at $1.50 a cord or less as compared with $3.25 a ton or more for coal, the advantage is noticeable.

Few operatives are needed in one of these small mills. These possibly may be secured from the surrounding farms and become valuable before they are seized with the desire to move constantly, the bane of the factory population. It is easier for the superintendent to secure the personal knowledge of his operatives necessary for success in management.[1] On account of the smaller cost of living, wages may be lower than in the mill centers. The rates of commission charged for selling the product are the same as the larger mills pay, or a neighboring mill takes the output. There is no complaint of freight discrimination between shippers in the same territory.

The only disproportionate expense, then, would seem to be cost of efficient management and superintendence. Often the superintend-

[1] See Chap. XI.

ents of these small mills are men of little education or theoretical knowledge, who have worked up from the position of operatives. They know well the practical side of spinning the few standard numbers, and little attempt is made to diversify product. They understand "managing help," but their lack of training unfits them for the management of larger and more complicated establishments.

Sometimes the superintendents are young men of education trained in larger mills who take these positions to make reputations. They throw all their energy into the work, knowing that a man of unusual ability and success will not long escape the eye of the managers of the larger mills. It has been increasingly difficult to secure men having that peculiar combination of qualities necessary for the large establishments. Custom has been against promoting a man in the same establishment, and every young man, no matter how obscure his mill, is spurred on by the possibility of securing one of the great prizes. Enthusiasm counts and often produces greater results than experience.

Further, a point in favor of the smaller mills has been the increasing difficulty in securing competent managers of the large mills. A large body of trained entrepreneurs, successful in the management of large enterprises, has not yet been developed. Men successful in the smaller business are not always proportionately so in the larger. When a large establishment loses a successful manager, trouble is often experienced in filling his place. The high grade of managing ability will become more common as a larger proportion of the population turns attention to business careers. Already there are individual managers able to meet successfully any competition, and their number grows larger. Meanwhile, the smaller establishment has paid the larger dividends.

This condition cannot endure, as many of the advantages enumerated are but temporary. The operatives show an increasing preference for the larger mills, a part perhaps of the general movement from country to town. The supply of wood is being exhausted, and as the larger mill has usually the more economical engines, the cost of coal will be proportionately greater in the smaller mills. The state spun in 1904 about 96.3 per cent of its cotton crop, and the local supply cannot much longer be depended upon.[1] The profits of the smaller mills, unless favored by exceptional local conditions, are not likely to continue so great.

[1] The larger crop of 1904 reduced the percentage of consumption for 1905 to 78.4 per cent.

## CHAPTER VI

### THE REAL FACTORY OPERATIVE

MUCH nonsense has been written and believed concerning social conditions in the South before the Civil War. The large planter with thousands of acres of land and hundreds of slaves on the one hand, and the "poor white trash," have been described as comprising the whole white population.

The Northern writers have not made sufficient investigation, and the few Southerners who have attempted to describe ante-bellum life have often done so sentimentally. Little attempt has been made to correct the prevalent impression, possibly from a desire to be considered as belonging, by descent at least, to the opulent aristocracy; possibly because the truth is not so interesting from a literary standpoint. A typical statement of a Northern writer is found in one of the widely read Chautauqua Series: ". . . there was no middle class in the South. The 'poor whites' were ignorant and degraded." [1]

A very slight study of conditions in the several states would have shown the inaccuracy of such sweeping statements. The white population of North Carolina in 1860, 629,942, representing perhaps 125,000 families, contained but 34,658 slaveholders, and these owned 331,059 slaves — an average of less than 10. Such a number could hardly raise the owners into the class of a feudal aristocracy. Moreover, of the total slaveholders, 18,316 owned less than 5, and 12,277 more from 6 to 20. Only 3321 owned from 20 to 50, 611 from 50 to 100, and 133 owned more than 100.[2] Of the 311 largest slaveholders, only 87 lived in the Western counties, though that section contained more than three fifths of the white population.

Land was, of course, held in larger tracts

---

[1] Beers, "Initial Studies in American Letters," 1895.
[2] U.S. Census 1860, *v*. Population, p. 351.

than in the North, but even in this respect, conditions in the South were by no means uniform. The average size of farms in North Carolina in 1860, including the large plantations in the East (some of which included much apparently valueless swamp land), and other great tracts of waste mountain land in the West, was 316.8 acres. This is to be compared with an average of 536 in Louisiana, 488 in South Carolina, and an average of 401.7 in the cotton states taken as a whole. In all 75,203 separate farms were reported.

Further, of all these farms 69.1 per cent. were of less than 100 acres, 28.7 per cent. more of 100 to 500, and only 2.2 per cent. of more than 500 acres.[1] The number for each of groups 20 to 50, 50 to 100, and 100 to 500 was almost the same, *i.e.* 20,882 for the first, 18,496 for the second, and 19,220 for the third. The number of farms above 1000 acres was exactly the same as the number of large slaveholders, 311.

[1] Consult an elaborate study of conditions, Von Halle, "Baumwollproduktion" (Leipzig, 1898).

The character and the life of the settlers of the state have already been described. In the East and the adjacent "pine barrens" were some large plantations, and an approach, perhaps, to a "poor white" class. In the extreme West, among the mountains, the inhabitants lived in 1860 the same primitive lives that their grandfathers had done, but in the great middle or Piedmont region different conditions prevailed.

Here was a strong, sturdy, middle class. The proportion of slaves to whites, 38 per cent., was smaller than in the East. Counting five persons to the white family, there were only 1.9 slaves for each group. The slaveholders in this section, often held only one, or a single family, though of course there were many who owned a larger number.

The Scotch-Irish and Germans who had come to this region were not desirous of escaping churches and schools. They sent back to Pennsylvania or to Germany for ministers, and, particularly among the Scotch-Irish, classical schools were established by the side of the churches.[1]

Clio's Nursery in Iredell, Zion-Parnassus in Rowan, and Dr. David Caldwell's School in Guilford were well known. Here young men, particularly those preparing for the ministry, were grounded in the classics, mathematics, and the strict Calvinistic theology which their preceptors had imbibed at Princeton. The statement of Fiske, that until just before the Revolution there was not a school, good or bad, in the province, is entirely untrue.[2]

Education was not, however, regarded as a right, nor as a necessity for every one. The interest of the church was foremost; and the dictum that "unsanctified learning has never been of any benefit to the church" was generally accepted. The efforts of the state to establish a satisfactory public school system and the partial success have already been mentioned.

[1] Consult Bernheim, "History of German Settlements and of the Lutheran Church in the Carolinas" (1872).

[2] Author, "Some Log Colleges in North Carolina," *Presbyterian Quarterly*, January, 1900.

Throughout this section the people were generally fairly industrious, "good-livers," to use a colloquial expression, though few became wealthy, even by the moderate standards prevailing before 1860. Adequate transportation facilities were developed slowly, and for a long time it seemed hardly worth while to raise more than could be consumed upon the farms.

The idea that manual labor was a disgrace had no foothold here. The more disagreeable kinds might, perhaps, be called "negro work," and a certain repugnance be felt for such occupations. But dozens of the older men have told me of working in the fields, plowing, hoeing, or gathering the crops, with the slaves.[1]

These slaves belonged to their fathers or were hired for the work. A landowner whose labor force was insufficient might hire the slaves of minor heirs under the protection of the courts; or he might secure the services of his less prosperous white neighbors. Whites thus

[1] For similar conditions in middle Georgia, read Joel Chandler Harris's stories.

employed by the smaller farmers ate at the family table and slept in the house, as they do to-day.

Many of the wealthier families in the Piedmont section, at least, owed the beginning of their fortunes to some artisan or small manufacturer who bought land and slaves with his profits. One ante-bellum United States senator of culture and distinction was descended from an iron worker; another from a hatter; a prominent political leader was the son of a cabinetmaker. Two prominent families, large enough, almost, to be called clans, are descended, respectively, from a blacksmith and a tailor.

Many men of influence were, or had been, merchants. There was no prejudice against trade. The position of the merchant was close to that of the lawyer and the doctor. It is true that all these classes were likely to be planters also. Every country doctor had a farm. Nearly every lawyer owned a plantation to which he expected to retire with advancing years. To live on a plantation combining *otium cum dignitate* was a well-nigh universal ambition.

The planter in this section did not despise the degrees by which he had gained his ambition. Indeed, it is not too much to say that the Piedmont section of North Carolina was more nearly a social democracy after 1840 than were the manufacturing sections of New England, where by that date there was a well-defined manufacturing aristocracy.

With the war and the ensuing disorder and demoralization, two opposing movements were apparent. Many of the small landowners lost their land, and became tenant farmers, some for a fixed money rent, but more for a share of the crop. Large plantations, however, were divided, and some of the negroes began to acquire land. By 1870 the number of farms had increased to 93,565, an increase of 18,362, compared with an increase of 18,240 from 1850 to 1860.[1]

[1] Census 1870.

The renter in some cases furnished his own stock, tools, seed, and labor, receiving in return two thirds or three fourths of the crop, though occasionally on rich alluvial lands two fifths were demanded for the land. Many more could furnish only the labor, and the landowner furnished the tools, stock, etc. While modified in individual cases, a rough understanding grew common, that in the division of a crop, one third was due to land, one third to labor, and the remainder to stock and tools.

Often, however, the renter was unable to sustain himself until the crop was gathered. In such cases, supplies must be "advanced" either by the landlord or by the country merchant. Hence there developed the system of crop liens and chattel mortgages, by which the farmer was enabled to mortgage his stock, and his growing, or even his unplanted, crops, to secure the necessities of life.[1]

Many small landowners also lacked capital,

---

[1] For an elaborate and valuable study of the tenant system see Hammond, "The Cotton Industry" (1897).

and were forced to seek advances from the merchant. No interest was charged upon such advances, but the merchant gained his profit by the higher prices charged for all articles. These prices have varied with the neighborhood, the character of the merchant, the reputation of the mortgagor, and sometimes with his necessity or ignorance. The general limits are perhaps between 10 per cent. and 50 per cent. above the regular cash prices. But as the account seldom stood so long as a year, and a large proportion of the purchases was made during the few months before the crops were gathered, the farmer, in debt to the merchant, has paid on a part or the whole of his working capital a rate of interest ranging from 25 per cent. to 200 per cent. a year. To some extent these conditions still exist.

Further, the opportunity to purchase on credit has always been a constant temptation to extravagance. The day of payment seems far away and the crop appears large in prospect. Though the usual stock of the country merchant is tawdry and uninviting, many things are purchased which are not strictly necessary. Before the crop is gathered the agreed advances are sometimes absorbed, and pinching economy may be necessary during the last few weeks of the season.

The merchant, also, has demanded a voice in determining the crops to be planted. Cotton and tobacco have been the favored staples. Both contain comparatively large value in small bulk. Neither is subject to deterioration to the same extent as other crops. A ready-cash market, at some price, is always present for either, and neither is liable to a total failure. To the grower there is also a gambler's chance of great profits.

The profit to be made on provisions has also influenced the merchant in his choice of the crops to be planted. So tons of Western bacon have been and are still sold in regions where hogs can be easily and cheaply raised. Western hay, corn, and flour are sold in districts admirably suited to the production of grass and grains.

When the crops are gathered, it is necessarily an indictable offense to sell without the mortgagee's consent. In fact, he is usually the purchaser, and the surplus, after accounts are settled, is paid to the farmer. Undoubtedly the price credited for the products is not always the highest that might be obtained in a free market. The grower cannot hold back his crop for a possible higher price, as the merchant naturally demands the early settlement of his account.

The net returns to the farmer of this wasteful system are of course small at best. A partial failure of his crop, or especially low prices growing out of generally excessive production, may render the payment of the merchant's accounts difficult or impossible. In such a case the merchant has the right to strip the farmer of his live stock and tools, but oftener the unpaid balance is carried over to the next year. The meaning of the expression, "Two years behind," is obvious. The arrears of several unprofitable years may reduce the farmer to a state bordering upon despair.

By no means all who purchase supplies on credit are so entirely in the power of the merchant, nor does the latter grow so rich as might be expected. In spite of all his precautions, a considerable proportion of his accounts are uncollectible, his assets are slow, and not easy to realize upon, if he is himself pressed. Often, too, he shows mercy to the unfortunate. During the period of depression following the panic of 1893, many merchants were crippled, since the price of all agricultural products was so low that the crops often would not pay the advances, and many farms were abandoned.

The Farmers' Alliance preached coöperative buying for cash and many members escaped from debt and have not been obliged to return. The high price of cotton in 1903–1905 freed thousands more. Except among those directly connected with manufacturing, one hears in the South little condemnation of the "Sully corner" of 1903–1904.[1] It is the general feeling

[1] The attempt of the farmers to secure a monopoly price for the crop of 1905 has been viewed in somewhat the same light.

that the farmer has had an unfair share of the burden of government; that he has been compelled to buy in the dearest and to sell in the cheapest market. Generally, speculative manipulation has been credited with keeping down the prices of agricultural products, at least until the crops have passed into the hands of middlemen and speculators.

In the Sully manipulation the farmer received much of the benefit, as the inflation began before the whole crop had left his hands. The higher prices brought freedom and hope to the Southern farmer, and stimulated trade in comforts and simple luxuries. A new buggy, a cabinet organ, a suite of furniture, a new kitchen stove, improved machinery for the farm, were some of the results. Perhaps a son or a daughter gained the coveted opportunity of a year at an academy or at college.

With these people, the small landowner and the tenant farmer from the surrounding country, the factory villages have been filled. The mountain counties have furnished comparatively few operatives to the North Carolina mills. The establishments in upper South Carolina have drawn a somewhat larger proportion of their operatives from the mountain people. No operatives were imported from abroad when the mills were built. There was then no urban population, hardly a village population.[1] The mills have been filled with a population coming from the soil, as was to a great extent the case in New England seventy or eighty years ago.

Here and there are, however, a few individuals who have traditions of culture and wealth. Descendants of men who possibly held high position in the state and nation now earn their bread at the spinning frames or at the looms. Ruined by the war in some cases, their fathers were unable to adjust themselves to changed conditions. The children have lacked the training to fit them for more responsible positions, and the mill furnishes a living.

These people are all Americans, and hundreds

[1] Chapter I.

could qualify as Sons or Daughters of the Revolution. They have lived simple, primitive lives upon the soil in a sparsely populated community. In 1900 only 36.6 per cent. of the land was in cultivation. They have lacked the stimulus arising from free association with the world. Even association with their neighbors was not easy.

The monthly or semi-monthly church services were attended with regularity. A day or two at the quarterly or semi-annual sessions of the Superior Court at the county seat brought many together. In fact "Tuesday of Court" sees perhaps the largest number of farmers in town next to "Circus Day." Political meetings are attended and the country store on Saturday afternoon has always been a meeting place. In the summer, when "crops are laid by" and farm work is slack, a neighbor's family comes to "spend the day." But of incidental social intercourse in the daily round of work there is little.

Facilities for education have been lacking,

and many are illiterate. This is particularly true of those growing up in the decade between 1860 and 1870. This ignorance does not necessarily imply that they are dangerous citizens. It is easy to make a fetish of the ability to read and write. Yet society may dread more the discontented literate than the ignorant farmer. Their ignorance has made them an obstacle to progress rather than a positive menace to the existing order.[1]

Their lack of knowledge has intensified their conservatism, and blinded them to possibilities of ultimate good at the cost of present discomfort; but nevertheless many of them think. Following the plow, they turn over in their minds the arguments heard at the political joint discussion, or the position of the lawyer or of the squire. Their thinking is not always clear nor logical, but often their sturdy common sense brings them to surprisingly logical conclusions. Illiteracy in a city slum and in a rural community are not identical dangers.

[1] Read Ingle, "Southern Side Lights" (1896), Ch. V.

The dangers of an illiterate mill population are yet to come in the state.

The term "poor white trash" applied so often by Northern writers to the mill population is almost unknown in Piedmont Carolina. During a residence of twenty-five years in that section, the writer heard the expression used by whites hardly a dozen times, and seldom by negroes. A servile white class does not exist. This fact cannot be stated too forcibly.

Speaking broadly, this stratum of the rural population — it is not a class — is an honest, self-respecting, law-abiding, God-fearing people. In many neighborhoods doors are not locked at night. As a result of whisky, personal encounters are not unknown, but the percentage of crime is low. Sexual immorality is not common. The people are poor, but the number of paupers is small. They are unprogressive, they fail to make the most of their opportunities, but they are not degraded. It is suspended or arrested development rather than degeneracy.[1]

[1] Page, "The Rebuilding of Old Commonwealths" (1902).

Intellect and ambition are dormant, rather than dead. Every year boys and girls, fired with the desire for learning, enter school or college. A large proportion of the graduating classes of many institutions comes from this stratum of the population. The college authorities can tell almost incredible stories of persistence in the face of difficulties. In spite of insufficient preparation, these boys and girls often stand high in their classes and become leaders in many departments of the college life.[1]

The motives for the migration to the mill are various. Some of the propelling forces have already been mentioned. Underlying all is the hope of bettering the general condition of the family, of receiving larger returns for the family labor. The hope of gaining at the mills better housing, better food, better clothing, together with the inarticulate social instinct, fills the factory tenements.

[1] For an instance, see *New York Evening Post*, June 5, 1905.

# THE REAL FACTORY OPERATIVE 115

Observation seems to divide those coming to the mills into five classes. First, is the honest man, ambitious for his children, who comes intending to work himself and hopeful of greater advantages for the education of the children, since the school term in the mill village is twice as long as in the country.

Next is the incapable or the shiftless, the man who lacks the mental qualifications or the moral steadfastness necessary for success in an independent capacity. He may work hard, but faulty judgment renders his efforts impotent. His condition at the mills is not likely to be worse and may be better.

A third class comprises those suffering from some physical disability, real or imaginary. They are, or fancy they are, incapacitated for the hard work of the farm and the necessary exposure to the weather. Perhaps the children are girls who cannot do the rough work in the fields.

The fourth class is composed of widows. Among these people, life insurance is not common. The death of the head of the family makes the cultivation of the farm a serious problem. If the family is to be kept together, the mill seems the only refuge.

There is another class, those who come to the mills with the deliberate intention of living a life of ease on the earnings of their children. Tired of the constant struggle upon the farms, they shift the burden upon younger shoulders. They discuss politics and neighborhood affairs around the store in the winter and in the shade in the summer. They carry the provisions from the store, and, perhaps, if the house is remote from the mill, take the dinner to their children. The epithet "tin-bucket toter" has been coined for them. This class is not large at first, but receives accessions from the first three classes, the members of which find it a difficult task to take up a new employment after age has taken away adaptability.

Whatever may have been the original motives of the immigrants, the general result is the great problem of child labor. The process and the extent to which the children assume the support of the family belong to the chapter on Child Labor, where that grave question will be discussed more fully.

# CHAPTER VII

### THE OPERATIVES AT WORK

NORTH CAROLINA, as has been previously stated, has a larger number of separate mills than any other state, though in production she is only third. The figures for 1904, the latest available, show 304 separate textile establishments [1] having 2,178,964 spindles and 48,612 looms, an average to the establishment (excluding the knitting mills) of 8285 spindles and 185 looms.[2] Mills with about this number of spindles are often found, while many mills have no looms at all. To follow a bale of cotton through a mill of average size and to study the processes and the workers who handle it, will give much aid toward a clear conception of the problem.

[1] Including forty-one knitting mills, and twelve small woolen mills the statistics of which are not separately given.
[2] The figures for 1905 which became available after the chapter was in type do not affect the general ratio.

The cotton is delivered at the mills in bales, packed in jute or cotton bagging, just as they came from the gins. The standard bale is 500 pounds, but the actual range is from 375 to 600 pounds, with more bales below than above the standard weight. There is no standard size, but the average is perhaps 30 × 48 × 54 inches.

The bales go first to the picker room, which is shut off from the rest of the mill by fireproof walls, or else is in a building entirely separate. Here the bagging and the ties (bands of strap iron encircling the bales) are removed. Handfuls are taken from several bales in turn and thrown into a bin in order to average as far as possible any differences in moisture, color, and length of fiber. The bales may have been raised upon different kinds of soil under varying climatic influences. The treatment at the gin and in storage may have been different. Some have been stored in a dry place while others have been exposed to the weather. Mixing in the Southern mills is not so important as in England or New England, where there is more variation in the raw material and where mixing different grades to secure stock at a given price is studied with care.[1]

The cotton thus mixed is fed to the "opener," which loosens the fibers that have been closely interlocked and compressed, and begins the work of knocking or blowing out the dust, trash, motes, and other foreign matter. The man in charge of opening the bales and of this machine is called the "opener." The task demands only strength and a minimum of intelligence. Sometimes the opener is a negro, and this is usually the only position inside the mill which one of that color may hold.

Next the cotton is fed into a "lapper," which continues the work of untangling the fibers, removing impurities, including broken ends or short lint. It is delivered in the form of "laps," which are sheets of batting of loose texture 36 to 45 inches wide and usually 48 yards long, weighing from 10 to 18 ounces the

[1] Nasmith, "Students Cotton Spinning," p. 90.

yard. For the purpose of further mixing and in order to equalize any differences in weight or thickness, a number of these laps, usually four, are superimposed and drawn out into one of the same weight as each of its constituent parts. This process is usually repeated, and occasionally a second time. A white man manages these machines.

The laps are now taken to the "cards," which continue the work of untangling the fibers and remove the impurities left by the previous machines. The fibers are rendered approximately parallel, and the cotton is delivered in the form of "sliver," which is simply a loose, untwisted cotton rope a little less than one inch in diameter and weighing yard for yard hardly a hundredth part as much as the lap. By an ingenious device the sliver is deposited coiled in cylindrical cans. One or two men, with the "card-room boss," who has general charge of all the processes thus far, can manage these machines in an 8000 spindle mill.

From four to eight cans of sliver go to each "drawing frame," of which the essential feature is pairs of rollers moving at unequal and increasing speed. The fibers are rendered parallel, and any inequality in the constituent slivers is made less important by combination with the others. The product delivered from the most rapid rollers is a single rope of practically the same weight as each of its constituent parts. This evening up is so important that the process is usually repeated twice. If six ends are fed each time into the machine, it is obvious that $6 \times 6 \times 6 = 216$ ends have been drawn into one. Two or three men or strong women will manage this process.

The sliver is now fed to the "slubber," which reduces the thickness and imparts a slight twist. The product is now "slubbing" and is wound upon large bobbins to be ready for the next process. The "intermediates" continue the attenuation and twisting, and the process is carried further by the "fine frames," which are almost duplicates of the intermedi-

## THE OPERATIVES AT WORK 123

ates. Two or three men manage the slubbers, the same number of men or women the intermediates, and four or five the "speeders," as the fine frames are often called. The stock, now become "roving," is ready for the final drawing out and twist imparted by the spinning frames proper, or by the "mules."

In the South, mules are little used, and practically all the yarn is spun upon ring frames. These are 36 to 39 inches wide, and have two sides. The length varies, but 27 feet is most common. This length contains, of the yarns most generally spun, numbers 16 to 30, 104 spindles on each side, or 208 to each frame.

So far the twist imparted has been only enough to keep the cotton together. Now the bobbins of roving are placed in creels, and the ends again run between pairs of rollers revolving at unequal speed. The spindles driven at high speed, from 5000 to 10,000 revolutions the minute, impart the twist, and by action of the "traveler" the resulting yarn is wound upon the bobbins. The high speed and the tension cause the threads to break frequently, and these must be twisted together.

Girls nearly all below the age of sixteen do this work, each one looking after from one to eight "sides," *i.e.* from 104 to 832 spindles. Eight sides has been the limit of economical operation in the South, as frequently several threads break at the same moment and that part of the machine is idle until they are mended.

Only the youngest beginners are confined to one side. The average in different mills lies between three and six. This varies with scarcity of operatives as well as with absolute skill. When labor is plenty, the number of sides allotted to each girl is smaller, as thus a nearer approach to production (the amount theoretically possible for each spindle to deliver) may be secured. As wages are paid by the side, each girl is naturally ambitious to run as many as possible. The work requires little physical strength, but a high degree of dexterity comes to the experienced spinner before she is advanced to the looms. In a mill where all machinery is first-class, when the raw material is of good quality and the atmospheric conditions are right, there is nothing to do for considerable intervals. On another day the threads break constantly and all possible nimbleness cannot keep all the spindles running. Constant watchfulness is always required.

The full bobbins are removed and empty ones placed in their stead by boys, "doffers," and the operation is called "doffing." They work exceedingly rapidly, but have long periods of rest. In all they work from 20 to 45 minutes in every hour. Often when they will not be needed before the closing time, they are dismissed before the other operatives, or occasionally are allowed to play out of doors until they are needed.[1] The mill of which we are speaking would have 40 machines, 80 sides, and 16 to 24 spinners. Nine or 10 doffers can keep the machines clear. In addition there

[1] Bulletin Bureau of Labor (U.S.), No. 52, p. 514.

will be an overseer of spinning and a section hand, both men.

The bobbins are necessarily wound with irregular tension, as the thread circles from bottom to top. To remedy this, the yarn from several bobbins is now wound regularly and smoothly, with no additional twist, upon a spool. Usually girls or women run the spoolers, and 8 to 12 will be required.

For single yarns the processes heretofore described are the same whether the yarn is to be woven on the premises or to be sold. If "ply" yarns are desired, 2 to 6 of the strands are twisted into a single cord, by special machinery managed by 5 or 6 men. But whether single or twisted, the processes through which the yarn now goes differ according to its destination. If it is to be sold, from 1000 to 2000 spools are taken to the "Denn warper," which draws them all into one great rope or skein, and knots or links them together to prevent tangling. It is then ready to be baled for the market. One man has charge of the ma-

## THE OPERATIVES AT WORK 127

chine. On the other hand, if the yarn is to be woven upon the premises, the threads designed for the warp (lengthwise), from 300 to 600 spools, go to the beam warper. Here they are wound upon cylindrical beams. Again one man has charge.

From 3 to 6 beams, depending upon the width, fineness, etc., of the cloth to be woven now go to the slasher to be "sized." The "ends" (separate threads), say 400 on each beam, now pass through a box containing starch, tallow, and sometimes other ingredients, which serve to stiffen and strengthen the yarn, and render it smooth. As they pass out they are drawn between heated cylinders, and the ends are wound upon a loom beam. One man only, with a little outside help for the heavy lifting, is required in this position.

The loom beams must now be "put into harness," as the arrangement of the heddles and reeds in the looms is called. Each separate end, sometimes more than 2000, must be drawn through an eye in the harness and a dent in the reed. Three girls do this work, called "drawing-in," which is probably more trying on eyesight and nerves than any other position in the mill. A recent invention promises to set free these workers.

These loom beams, with the ends drawn into harness, are now adjusted in the looms. The filling (threads running across the cloth) comes directly from the frames on bobbins ready to be placed in the shuttles. The operatives are adults, men and women. Their duties are to keep the ends mended, and fresh bobbins in the shuttles. Some strength and judgment is required, as the loom is a complicated machine. The number of looms which can be managed by a single weaver varies with the quality, weight, width, and color of the cloth, the style of the loom, and also with the skill, strength, and natural or inherited aptitude of the weaver himself. Occasionally a weaver will manage 8 common looms, excellent weavers have 6, a greater number has 4, and the younger and more inexperienced have 2.

With the automatic loom, which throws out the empty bobbin and takes a full one from a creel, an operative can manage from 12 to 24. In our mill of 8285 spindles, the number of regular looms necessary to consume the yarn varies from 200 to 250, depending upon the fineness of the yarn and the goods woven. There will be 45 to 60 operatives, a "weave boss," and 2 "loom fixers." Payment is by the cut of 40 to 60 yards.

The cloth as woven is wound upon a beam holding several cuts. The cloth on a number of beams is sewed into a strip and passes through a machine variously known as a "brusher," "shearer," or "calendar." This shears off the loose threads, emery wheels grind off the rough places, and after the cloth has passed through a steam jet, heated rollers iron it smoothly. Next the cloth goes to the "folder," which makes the bolts seen in the shops. After stamping and baling, the cloth is ready for the market. Two or three men have charge of these three processes.

From the foregoing account it appears that the labor force employed in the average North Carolina cotton mill which spins only is as follows: —

| | |
|---|---|
| I. 1 Superintendent. | IX. 4–5 Speeder hands. |
| II. 1 Card-room boss. | X. 1 Overseer of spinning. |
| III. 1 Opener. | XI. 2 Section hands. |
| IV. 1 Picker hand. | XII. 16–24 Spinners (girls). |
| V. 1–2 Card hands. | XIII. 8–10 Doffers (boys). |
| VI. 2–3 Draw-frame hands. | XIV. 8–12 Spoolers. |
| VII. 2–3 Slubber hands. | XV. (4–6 Twisters, if ply yarn is desired). |
| VIII. 2–3 Intermediate hands. | XVI. 1 Warper (man). |

In addition there will be: —

| | |
|---|---|
| XVII. 1 Band boy. | XX. 1 Baler. |
| XVIII. 2–3 Sweepers (old men). | XXI. 1 Engineer. |
| | XXII. 1 Fireman. |
| XIX. 1 Oiler and bander. | XXIII. 2–4 Truckmen. |

Say 40 to 50 adults, and 28 to 40 children, as occasionally a boy or girl below the age of 16 may work at the draw frames, spoolers, or twisters. The percentage of children ranges from 35 to 45.

If the mill has looms and weaves its yarn into cloth, the warper (XVI) will be omitted, and we have in addition: —

XXIV. 1 Filler (man).
XXV. 1 Spooler to warper (woman).
XXVI. 1 Beam warper (man).
XXVII. 1 Slasher tender (man).
XXVIII. 3 Drawing-in girls.
XXIX. 1 Weave boss (man).
XXX. 2 Section hands (men).
XXXI. 45–60 Weavers (men and women).
XXXII. 1 Calendar (man).
XXXIII. 1 Folder (man).

Fifty-five to 70 additional employees will be needed, practically all above 16 years of age. The drawing-in girls and an occasional weaver may be younger. It is obvious that the proportion of children in a yarn mill is much larger than in a cloth mill. In a mill where no spinning is done, as in many mills around Philadelphia, the number of children will be small, almost negligible; but as the proportion of spindles grows, the number of children grows with it.

This is particularly true when the spinning is done upon ring frames. It is not so true where the mules are extensively used. The "mule," which is to-day simply an elaboration of Crompton's original invention, has the spindles attached to a movable carriage which travels away from the rollers which deliver the roving, and in this progress receives its twist equally. The yarn is wound evenly upon the spindles as the carriage returns. The regularity of tension causes less breaking of the yarn. Much finer yarn can be spun upon mules than upon frames for this reason. Numbers 60 to 100 is the limit upon frames, while number 500 may be spun upon mules.[1] The yarn is also softer and for some purposes is indispensable. Only men or exceptionally strong women can operate these machines, sometimes containing 1500 spindles, and a high degree of skill is necessary. The process is more expensive and the product per spindle is less. So far the Southern mills have been occupied chiefly with the lower numbers, and few mules are in operation. In New England

---

[1] In the notation of yarn the unit is the relation of the "hank" of 840 yards to the pound. Number 20 yarn means that 20 hanks each of 840 yards will be required to weigh a pound; number 36, that 36 hanks will weigh a pound, and so on. It is obvious that the lower numbers are the coarser.

a large proportion of the spindles are upon mules (4,477,199, compared with 8,373,788 on frames [1]), and this has something to do with the proportion of children employed, entirely regardless of any legal enactments. Where a class of operatives cannot be used profitably, economic law alone will prevent its employment.

The hours of labor in the North Carolina mills are long. Before the passage of the act of 1903, limiting the number of working hours in a week to 66, the length ranged from 63 to 75, with the average close to 69. In the early days of manufacturing there was no objection to these long hours. The length of the day in the fields during the summer was much longer. "From sun-up to sun-down" was a rough method of measuring the working day of the unskilled laborer. Since the work in the mills required much less muscular exertion, the hours were not considered excessive. The fact that this longer day in the fields was in force for only

[1] Census 1900, Bulletin 215.

a part of the year, was not considered. Since the operatives did not complain, and, in fact, petitioned against a change, the public was not inclined to interfere. The feeling that a contract for wages and hours is a matter for the parties immediately concerned, was strong.

With the agitation for shorter hours and an age limit, came strong opposition from all parties directly concerned. The legislature of 1901, however, would have passed a bill, but for an agreement signed by most of the mills, limiting the hours of labor to 66, and the minimum age of operatives to 12 years. This agreement was not faithfully kept, and the legislature of 1903 enacted the present law.[1]

Since the passage of the act limiting the week to 66 hours, the following scheme has been followed. The day operatives enter the mill at six in the morning and work 12 hours, with an intermission of 30 to 45 minutes for dinner at noon, on 5 days of the week.

[1] For further discussion, see the chapter on Child Labor, p. 219.

On Saturday they work from six until twelve. The night operatives work from six-thirty or six-forty-five until six in the morning, with an intermission of 15 minutes at midnight. On Saturday night, of course, work stops at twelve.

Night work has been almost universal, particularly in the spinning department, though it is now decreasing. The spinning frames have been run 22 or 23 hours in every day. This has been done to keep up with orders and to wear out the machinery. Night work is always inferior to that done by day. The best operatives will not usually work at night, and many of those who do, cannot or will not take sufficient sleep during the day. The younger operatives are more careless and inefficient. Not so much is accomplished, though the wages per hour are always higher. During especially profitable periods mills have sometimes run only 5 nights while paying for 6.

But even if the percentage of profit on the night work is much smaller, it may be counted as profit in the total production of a machine. Improvements in machinery have come so rapidly that a machine can seldom be run until it is no longer capable of effective work. While it is still in good condition it is supplanted by an improved pattern, which the constant competition of the new mills forces the older ones to install.

The used machinery, though in good order and capable of good work, must be sold for a small fraction of its cost, or else discarded outright and sent to the scrap heap. By running it both day and night a greater proportion of the total effective productivity may be utilized. Many shrewd operators do night work for this reason. The social disadvantages appeal so strongly to others, that only when the lure of profits is tempting, do they yield to the pressure, while some mills have never run at night at all.

# CHAPTER VIII

## WAGES AND COST OF LIVING

IN studying the economic condition of a people, it is necessary, before pronouncing judgment, to consider status, environment, and inherited customs, as well as purely material considerations. This is difficult, since the temptation is unconsciously presented to transfer one's own standard of comfort in his own station, or in his own locality, to the locality to be studied, and measure the condition of a population by it absolutely.

The fact that necessities in one section may be absolutely superfluous in another, is disregarded. An item which forms a considerable part of the budget in one place may be lacking entirely in another. Food, dress, etc., are regarded differently, owing to training, habits, and former manner of life. Comparisons between workmen in different localities are rendered valueless by this lack of discrimination. The necessity of taking all these things into account is a part of elementary economics in its relation to life, but it is constantly neglected.

So, in studying the economic condition of the factory operatives of North Carolina, we must take into consideration, as well as their wage, the demands made upon that wage by the climate, the cost of subsistence, and the prevailing ideas of comfort and luxury not only in the operative class, but among the population in general. Care must also be taken to study the problem broadly and not emphasize some comparatively insignificant part of it, to the neglect of more important considerations.

The factory population was born upon the farms, or is only one generation removed. The operatives have come to the mill with generations of fixed rural habits behind them, and necessarily are greatly influenced by their past. The problem of adjustment is more than to adjust the younger generation.

Their standard of life was simple if not low.

There was food enough to satisfy hunger; there was clothing enough to give warmth; there was fuel enough for the great fireplace, though to keep the houses thoroughly warm in every part, while the wind whistled through the chinks and crevices, was almost impossible. The snow sometimes sifted through, but there was covering for the feather beds. Judged by the standard of the city dweller, their lot was intolerable, but they did not know it.

Luxuries bought with money were few. Many farmers who live in comparative comfort do not handle $200 in cash in a year. The chief money crops are a few hundred pounds of tobacco, or a few bales of cotton. The greater part of the food supply is raised upon the farms. Chickens and eggs or vegetables may be exchanged for sugar and coffee, but there has been no development of the trucking industry in the Piedmont section.

When they come to the mills, they live generally in the houses built by the corporation, though some employers urge their operatives to buy or build homes of their own. These tenement houses contain two to six rooms, rarely more, and more than one family seldom occupies a house. They are detached frame structures, built upon brick pillars. The rooms are either plastered or finished in the natural pine. They are fitted with open fireplaces in the larger rooms, and perhaps stove flues in the smaller. The lot is usually about half an acre, and the frontage 100 feet, though occasionally not more than 75.

When the mill is built in the woods, the trees are left for shade, but oftener some bare, worn-out hillside is the site of the village. Little grading is done, and the supporting pillars on one side may be six feet higher than on the other, giving the house the appearance of being perched upon stilts. One magazine editor on a tour through the mill region was more impressed by this than any other sight. The fact that the houses had no cellars, seemed to him proof of squalor and wretchedness.

As a matter of fact, few Southern houses have cellars. In some sections it is difficult to keep out the water, the climate does not make a cellar necessary for storage, and few houses have furnaces. The most expensive houses in a small Southern town will be built entirely above the ground, on brick or stone pillars, though usually these are connected by lattice work.

These mill houses have no running water, as few villages have a water system. Water is generally secured from wells, though occasionally from hydrants. The privy on the lot may be an unpleasant feature. A mill village is often monotonous. The general style of the houses and the colors are similar. Often streets and sidewalks are neglected, and the whole atmosphere may be depressing.

These houses are usually rented at a flat rate per room, regardless of desirability of location. A four-room house will cost 50 cents to $1 per room per month, *i.e.* $2 to $4. The rate is seldom higher, and at some mills no rent at all is charged. Houses are allotted in order of application, subject to an understood rule that a house must furnish one operative for each room, or at least two operatives for three rooms. When the demand is pressing a small family, or one from which few members work in the mill, may be unable to secure a large house.

So far as convenience and comfort are concerned, these houses, when new at least, are superior to those in which the operatives lived in the country. Of the houses themselves, there is little complaint. Often the location, the comparative crowding, the lack of shade, are causes of regret to the tenants.

The houses are often somewhat bare of furniture. The newcomers bring little to the mills, and that of the rudest description, but additions are soon made. The more prosperous have a parlor, with a center table on which lies a large family Bible and a few expensive books bought from agents. Bright lithographs, or a perforated cardboard motto, "God Bless Our Home," are upon the walls.

Perhaps the "company bed," with its huge embroidered pillow shams, stands in one corner.

The operatives dress well, or at least have good clothes for Sundays and holidays. In the mills these are not worn, any more than a machinist would wear his best in the grime and oil of the shop. During at least half the year the children are barefoot, as they were in the country. The clothing of a boy in summer is limited to the same two necessary garments worn by his country cousin. The men go to their work in garments white with lint, and work in their shirt sleeves. The girls have working clothes which catch as little of this lint as possible, but all have better clothes than they wear daily. In fact, there is a decided tendency to extravagance in dress, particularly among the girls. They are students of style and follow the fashions observed on the streets, generally emphasizing colors and modes. Many overdress, and wear too many and too harshly contrasting colors, and too many ornaments.

Though not always well chosen, the food of the operatives is abundant. In the sparsely settled country districts, butcher's meat was uncommon. Beef was had only at infrequent intervals when a farmer killed an animal and supplied his neighbors. Mutton also was rarely seen, since the number of worthless curs owned both by negroes and whites makes the raising of sheep hazardous. There are few fish in the streams in the middle section, and salt mackerel was almost the only kind known to the farmer's table.

Pork was the chief dependence, fresh at "hog-killing time," which comes after the weather has grown cold enough to insure preservation and salted for the remainder of the year. Occasionally rabbits, squirrels, and quail were upon the table, but the staple meats were pork and chicken.

The list of vegetables is not long: cabbage, corn in season, beans, potatoes, both white and sweet, tomatoes, cucumbers, and onions, with fruit and berries for pies make up the list.

For winter use cabbage was banked or made into kraut, berries and fruits were dried, and the other vegetables were kept so long as possible. As some of these do not keep well, sometimes for months in the late winter and early spring, the usual food was salt pork three times a day, varied occasionally by eggs or chicken, with few or no vegetables.

After the move to town, there is little change in the menu for a time, but gradually additions are made. The flour in the country was often ground at a buhr mill and was not perfectly white. The family begins to use fine roller flour, though it may be tinged yellow with soda. Fresh fish are bought. Canned fruits and vegetables form an agreeable addition, and are used extensively. Pickles and preserves are bought in large quantities. The operatives are large consumers of fruit and vegetables out of season, though they may put the strawberries into a pie.

With the staples there is less change. The cooking was generally bad in the country and it remains bad in town. The mother had the help of her daughters there, while in town she may be so hurried that she has not time to prepare the food properly. The frying pan is almost universal. Ham, bacon, sausage, chicken, eggs, come to the table swimming in grease. The best steaks are bought, but they are cut thin and fried to a crisp. Soup is rarely, almost never, seen. The vegetables are often boiled with the bacon and are greasy. The pie crust may be soggy, and the huge, yellow soda biscuits as well. Molasses in large quantities is consumed both in the country and in the town.

Conditions in the second generation improve little. The more expensive foods have become a part of the standard of life, but the cooking may not be better. The girl who has worked in the mill from childhood until her marriage can know little of housekeeping, and very often is unable to gain knowledge from experiment. Her utensils are better, however, and some learn.

In the country the open-air life aided the stomach to perform its difficult task. When the same food is eaten by those who have spent the whole day indoors, the stomach revolts. The faces and carriage of many give evidence of malnutrition. The craving for pickles and sweets is gratified excessively with disastrous results. Of course, this description is not universally applicable. In many cases the food is well chosen and well prepared, but oftener conditions are as stated above.

The factory population is obviously not stinted, so far as the kind and quantity of food is concerned. Other inhabitants of the town, clerks, mechanics, and artisans, spend hardly so much per capita, though the results gained may be greater.

How can the operatives spend so much upon food? First, because the item of rent is almost eliminated, forming, as it does, only a tenth or a fifteenth of the total income; second, the amount spent for dramatic and musical performances, books, magazines, and other cultural expenses is small; third, because little is saved. Though the family may have come with the expectation of accumulating enough to pay off a mortgage or to buy a farm, in rare instances is any considerable portion of the income laid by.

In the Appendix will be found a table of average wages paid for the different operations.[1] The family wages depend upon the number, age, and efficiency of the workers. A few small unskilled families make less than $10 a week, while fewer make so much as $30. The families accustomed to the irregularity of farm work, when they first come to the mill do not work so regularly as those whose rural life has become a memory. The average family wage is between $10 and $15, since the families by a natural process of selection are large. There is, of course, no car fare to pay.

In a typical, prosperous mill family the weekly wages list was as follows: father, $4.50; daughter, twenty-one years, $4.50; son, nine-

[1] Appendix A.

teen, $5.40; daughter, sixteen, $3.60; son, fourteen, $3; total $21, from which $1.50 weekly only was paid for rent. Three years before the amounts received were, respectively, $4.50, $3.60, $4.50, $2.40, $2.40, a total of $17.40.

Another family consisted of a widow and four children. The oldest, a girl of twenty, earned about $6.50 a week, while a sister of eighteen and a brother of seventeen earned about $4.25 each. The youngest daughter, a child of fourteen, had not worked regularly in the mill, but had occasionally assisted her sisters, coming and going when she pleased. Here, of total weekly wages of $15, $1.25 went for rent.

These are, of course, the more prosperous families. One widow whose two children together earned less than $5 a week managed to exist by taking two boarders who paid perhaps $3 a week additional. Obviously the margin was very narrow, and sickness would reduce the family to want.

In spinning and weaving, a general average

of wages can be secured with difficulty, since the pay depends upon the number of machines operated, or upon the quantity of goods produced. Where there are such wide differences of skill, the variations in pay are likewise great. The late Professor Mayo-Smith clearly showed that any table of average wages is practically meaningless where much variation exists.

In spinning, for example, the pay depends upon the number of "sides" tended. A learner may have for a few weeks only one or two sides. (The stories of children working for 10 to 20 cents a day have this much foundation.) A few months later the same girl will be tending four sides, with a corresponding increase in wages, and good spinners tend more.

Weaving is paid by the " cut," the length of which varies with the goods, the market for which it is designed, etc. An unskilled weaver with few looms may make no more than $2.50 a week, while an expert may receive $9, which is close to the maximum.

Comparison with the average wages paid in

New England is difficult, since the work is often not comparable. The expert male mule spinner upon fine yarns cannot justly be compared with the raw girl producing coarse numbers upon ring frames. The skilled weaver upon "fancy" cloth belongs to a different division from the producer of coarse gray sheeting. As yet the larger number of the mills in North Carolina is occupied with coarse goods on which great skill is not required, while the wages of the highly skilled operatives of New England raise the average there. The individual earnings in New England are undoubtedly larger than in North Carolina. The Census reports gave the median wage instead of the average, *i.e.* the wage standing midway between the highest and the lowest. Beamers and slasher tenders received in New England $10.50 weekly compared with $6 in the South as a whole; card hands $7 compared with $4.50; male weavers $7.50 compared with $4.50; female spinners $6 compared with $3. The great difference for the last operation is partially due to the excessive number of young and untrained girls in the Southern mills.[1]

For example, spinners in Fall River were found tending 10 sides of 112 spindles each, while the average was 8. In New Bedford some operatives had 1200 spindles.[2] In few mills was the average number of sides to the spinner less than 6. In North Carolina the frames are somewhat shorter, few girls manage 800 spindles, and the average is closer to 400. Undoubtedly, however, the rate paid was less in North Carolina, though not so much less as the wages received would seem to show.

A more important question for the manufacturer is that of comparative wages per *unit of efficiency*. That is, does the operative in North Carolina receive a smaller share of the total productivity imputable to labor? To settle the question, a careful study must be made

---

[1] Census Bulletin 215. Wages in North Carolina have risen decidedly since 1900, however, and in many places the rate for spinning particularly is as high as in New England.
[2] Young, *op. cit.*

## WAGES AND COST OF LIVING    153

of the comparative skill of workers, the amount of material wasted, and the state of the machinery both at the beginning and at the end of the period. An operative who gets from a machine a large percentage of its theoretically possible productivity may be cheaper at a high wage than one who gets much less per machine for similar work. The only trustworthy answer could be gained by comparison of results of mills under the same general management, but situated in different sections.

That careful English observer, Mr. T. M. Young, already mentioned, has the following to say of Southern wages: [1] —

"Wages are unquestionably very much lower and the truck system [2] is almost universal, but whether the cost per unit of efficiency is greater in the South than in the North is hard to say. But for the automatic loom, the North would, I think, have the advantage. Perhaps the truth is that in some parts of the South where the industry has been longest

---

[1] Young, "The American Cotton Industry" (1903).
[2] This is not true in North Carolina. Wages are paid in cash almost invariably.

established, and a generation has been trained to the work, Southern labor is actually as well as nominally cheaper than Northern; whilst in other districts, where many mills have sprung up all at once amongst a sparse rural population, wholly untrained, the Southern labor at present procurable is really dearer than the Northern. In any case I do not think that really cheaper labor can be counted on as a permanent advantage for the Southern cotton mills."

In support of this judgment he cites weavers working side by side in a North Carolina mill, some of whom were producing barely three fourths as much to the machine as others. As that machine was expensive, the inferior labor was higher priced, though receiving the same per unit of product; and the fact of the employment of inferior labor was obviously due to the scarcity of more efficient workers. He found, also, that both spindles and looms in the Southern mills ran more slowly than in the Northern.

This does not mean that the labor is permanently inferior, but the demand has been so great that an excessive number of totally

## WAGES AND COST OF LIVING 155

unskilled workers has been brought into the industry, and a long period is necessary to develop a class. Already individuals equal to any work may be found, and in some localities a considerable number. The longer hours, too, allow somewhat larger production. Mr. Young found that a mill in Massachusetts produced in a week of 58 hours 1.35 pounds of yarn to the spindle, while a branch mill in the South produced in a week of $67\frac{1}{4}$ hours 1.42 pounds of the same yarn.

From the standpoint of the laborer, the important consideration is the purchasing power of the wage, rather than its nominal size. A low wage in money may in reality be high on account of the smaller demands upon that wage.

The tables given in the Appendix [1] cannot of course show absolutely the relative well-being procured by the expenditure of a given sum, but they have some force. The prices given for Massachusetts are from the Report of the Massa-

[1] Appendix B.

chusetts Bureau of Labor, and are for 1897 and 1902, the latest procurable. The figures for North Carolina were obtained by averaging figures obtained from different mill towns.

The third table [1] represents the prevailing prices in two similar towns with practically the same industries, one in Piedmont Carolina, the other in the Connecticut valley. The date is April, 1904.

From the tables it is evident that the purchasing power of the dollar differs decidedly. Flour costs nearly the same, but the larger consumption of the cheaper corn meal (which is extensively used from choice by all classes in the South) makes the cost of bread less in North Carolina. Groceries generally show a slight advantage in favor of New England. The lower price of coffee in North Carolina means, of course, the use of a lower grade.

In meats there is a decided advantage in favor of North Carolina. In some cases the prices are little more than half. The general

[1] Appendix C.

quality is lower, to be sure, since it is nearly all from animals slaughtered in the neighborhood; but it is the best in the market, the same the employer eats. The consumption of pork is greater from habit, and it is cheaper, pound for pound, than beef. The same remarks apply to butter as to meat. The quality is not uniform, but the operatives eat the best the market affords. Eggs are cheaper in North Carolina, though slowly rising in price. For years, except around the holidays, ten cents a dozen was a standard price. During the summer vegetables are ridiculously cheap. There is usually land enough around the tenement for a garden upon which a family may raise a part of the supply, but oftener it is purchased from the store or from the farmers' wagons.

Fuel is much cheaper nominally and actually. Both wood and coal cost less, and less is required during the cold season. As all the cooking is done by wood, fire is kept in the stove only while meals are being prepared, and a decided saving is effected.

Dry goods for equal qualities show a slight advantage in favor of New England. The wide variation of quality in woolen goods prevents exact comparison. Less clothing is required for comfort, however, in the South, and the extensive substitution of cotton allowed by the climate makes considerable saving, though the North Carolina operatives may be clothed decently and suitably.

In the matter of rent, the North Carolina operative has a great advantage compared with Massachusetts particularly. In some parts of New Hampshire and in the Connecticut town already mentioned the difference is not so marked. In the latter town, however, the dwelling place is not a separate house, but only one half to one sixth of a tenement, and the surroundings are neither so healthful nor so pleasant as in the North Carolina town compared. It may, perhaps, be objected that a comparison of city prices in Massachusetts with those in semi-rural communities in North Carolina is not fair. Wages are compared, however, and the North Carolina mills are nearly all in small towns. If living is cheaper, it is to the advantage of the operative, and does not prevent a fair comparison of general conditions.

A comparison of wages paid in other occupations in North Carolina is not to the disadvantage of the factory worker. An average of the highest monthly wages paid to women in agriculture is $11.54, while that average in the mills is $27.04. The average wages paid to children in the mills was $10.66, while on the farms it was only $5.50. The average wages paid to able-bodied laborers on the railroads was 85 cents a day.[1]

A proof that wages do not bear hard upon the minimum of subsistence is the fact that the operatives have been induced to leave the farms, and that there is land to which they may return, and secure a subsistence no matter how unskillfully the labor is applied. These people are not city dwellers, to whom the country

[1] Report, Bureau of Labor and Printing, 1904.

is unknown. They have come from the farms and have not lost their connection with the rural community.

In the spring of 1904, when cotton was abnormally high, the possibility of closing was announced to the operatives in Concord. Though the mills were not closed, the census taken for school purposes in September showed the loss of nearly 1000 persons, 10 per cent. of the population. A large majority of them had gone back to the farms to raise a crop of cotton. The low price of cotton drove them from the farms; the high price lured them back, but not to stay. Nearly all of them returned to the mills, after selling their crops at a much lower price than they had hoped to receive.

But the importance of this point must be emphasized. The rate of wages must be influenced by the presence of the land. The demands of industry are encroaching perceptibly upon the supply of farm labor available either for wages or for a share of the crops. The high price of agricultural products in 1904–1905 compelled a decided advance of wages in the cotton mills.

One manufacturer reports that the amount of his weekly pay roll has increased twenty-five per cent. since the beginning of 1905.[1] While the increase in the rate of wages has perhaps not been so great in the state as a whole, the growing scarcity of labor is leading many employers to consider means of attracting foreign immigration.

[1] Letter, April, 1906.

## CHAPTER IX

#### SOCIAL LIFE AND AGENCIES FOR SOCIAL BETTERMENT

LIFE on the farms is lonely. Sometimes for days no outsider is seen, except the casual traveler along the roads, who halts to talk a few moments. It is not surprising that the farmer is always willing to stop and sit upon the fence to learn the news, or to make an excuse to go to the country store. Notices of political meetings, an unexpected church service, a debate at the schoolhouse, and the like are spread by word of mouth. Each listener then continues to "put out the word" until the whole neighborhood is notified.

But such occasions are comparatively infrequent. The family is thrown almost entirely upon its own scanty resources. Books are few, and many families receive no newspaper.

Perhaps the children turn over and over again their school books until they know them almost word for word. Many of these children, in spite of the short school term and, frequently, of inefficient teachers, at sixteen are as capable as the city children who have attended for terms twice as long.

In the country the influence of the church is strong. There is now more Puritanism in the South than remains in New England. The Scotch-Irish Presbyterian held a stern doctrine, and Drumtochties are still to be found. Though the Methodists and Baptists have added an emotionalism foreign to the old Presbyterian temperament, their rules of conduct are no less strict. The instance of a good woman who gave to her neighbor's children the nuts from a tree in her yard, provided they would promise not to crack them on Sunday, is not an isolated survival. There are thousands like her. Practically every resident of a rural community is connected with some religious denomination.

Naturally the loneliness, and the strictness of the imposed code, make a serious population. There is little spontaneity or hilarity in the ordinary rural gathering. Men, women, and children are usually sedate and quiet, almost grave. There are instances of reaction, of course, individuals who break away, whose spirits cannot be confined nor restrained. When they have offended against one social or ethical convention, they may be ready to offend in all. But viewing them in the large, these small farmers exhibit little of the "joy of life."

They come to the mill, and begin a new life. In the monotonous little mill village, they find excitement, and their starved social natures are gratified. The mothers talk from the windows or the piazzas with the neighbors as all go about their household tasks. Those in the mill are associated with their fellows, even though the noise and the nature of the work forbid much conversation. The children not in the mill have playmates.

Though the dwellers in the factory village

meet thus incidentally or at the church services, there is little formal social intercourse. The young men and the young women are together on Saturday afternoons and upon Sunday. Some take walks, and all the horses in the livery stables are engaged for drives on Sunday afternoons. Parties are rare. After the long day's work, bedtime must come early. The elders frown upon both dancing and cards. In fact, much of the old Puritanism which holds that any amusement whatever is wrong, or at least of doubtful propriety, still survives. This attitude may persist, though the church services are no longer so scrupulously attended.

This is particularly true in the rural mills, where the whole village is dependent upon the enterprise. Many managers are strict moralists of an old-fashioned type, who hold themselves responsible for the conduct of their operatives and attempt to control it in some particulars. Only those families who are willing to observe the regulations are kept, and the village often takes a somewhat austere tone.

It is easy to take cognizance of the conduct of the operatives outside the mill. In the factory village there can be little separation of private and industrial life. The houses are close together. Every one knows every one else, and can estimate the family income to a dollar a week. Every action almost is known, and fancied or real immorality cannot be long concealed.

The moral atmosphere of the mill settlements varies. At some old mills which have gotten a bad name through the carelessness or indifference of the manager or superintendent, and to which self-respecting operatives refuse to go, conditions are undoubtedly bad. The best operatives will not go where the tenements are bad, and sometimes short-sighted economic policy leads the managers to take the operatives that will come, instead of improving conditions. The less scrupulous operatives naturally tend to gather at the mills in the larger towns, where less supervision can be exercised.

At a great majority of the mills the atmosphere is clean. The testimony of unprejudiced Northern observers is quoted elsewhere, as have been examples of the scrupulousness of the operatives in money matters. As a result of whisky illicitly procured, sexual jealousy, or a hasty word, personal encounters sometimes occur, but they are seldom serious. The operatives marry young, and sexual immorality is not common.

The sweeping indictment against the chastity of the mill girls made in a recent novel, purporting to describe life in the Southern mills, is cruelly unjust.[1] There are many individual cases of unchastity, of course. No claim of universal purity is made for them. The conditions of the work, the crowding, the necessarily close association with the men would supposedly have a tendency to diminish maidenly reserve. There is vulgar conversation sometimes, and perhaps occasional profanity;

[1] Van Vorst, "Amanda of the Mill" (1905). As a picture of conditions, the book is untrue, and shows either ignorance, or perversion of facts for literary effect.

but the overwhelming majority of the factory girls in the North Carolina mills are virtuous, and follow the right so far as they know it.

In many mills the girls themselves make up an unofficial committee for the protection of social purity, and allow no offender to stay. In one mill where any deviation was punished by the discharge from the mill of the whole family and eviction from the tenement, the necessity for the infliction of the penalty did not arise for more than five years. Yet the population had changed considerably during that period, for the operatives can and will move upon an hour's notice.

This readiness to move is a symptom of social unrest arising from lack of adjustment to environment. The family, when it comes to the mill, may regret the loss of some of the advantages of the country it has left, and moves to another village in search of more satisfactory surroundings. Failing, the fault is attributed to the particular mill instead of the life itself, and other removals are made until the desire for constant change becomes chronic. The most trivial happening serves as an excuse. One man who moved his family sixteen times within five years gave as his reason for one of these transfers the fact that the wages of one of his children had been reduced ten cents a day. Such a record is fortunately unusual. On the other hand, a mill manager made special gifts a few years ago to a number of operatives who had worked in his mill more than twenty-five years. Yet the mill families often change. In a row of seven houses in one town only two had been occupied by the same tenants for more than a year, and some of the others had been occupied by three or four families in turn during that period. These tenements were in a large town, however, and were unsatisfactory from several standpoints.

Naturally such migratory families add little or nothing to the strength of a community, and are almost certainly a source of weakness. They send down no roots into the soil, form no real connections with their fellows about them.

The children lose whatever opportunities of education they might have had, and the church connections of the family are weakened or loosed altogether. Such a family does not make up a part of the public opinion, nor does it feel its full restraining force. The result is often a weakening and lowering of ethical standards. Such frequent removals bring loss of household comforts, which are destroyed or thrown away. A considerable portion of the family earnings is spent in transportation. Though the mill to which they go often advances the cost of removal, naturally the whole or a part of the cost is deducted from the future wages.

No matter how dissatisfied such a family may be with its surroundings, and no matter how vain seems the search for a satisfactory location, it seldom, almost never, returns to the farm. The reason is perhaps complex. The greater apparent earnings, even though the family wages may be spent before they are paid, make possible the enjoyment of certain comforts and luxuries unknown upon the farm; to return would seem a confession of failure, and they are still sensitive to the opinion of the neighborhood they left. Greater than all else is the morbid craving for excitement and change — a feeling analogous to that which keeps certain sections of the city overcrowded.

The formal agencies for social betterment at the North Carolina mills are the church and the school. Both are subsidized by the corporation. In nearly every mill community outside of those incorporated towns which maintain an efficient school system, the mill erects a school building. A school is maintained entirely at the expense of the corporation, or the short term of the public school is extended to six or even ten months by an appropriation from the mill treasury. Sometimes in towns which have a satisfactory school system the mill builds schoolhouses near the mills for the convenience of the operatives, and the town maintains the schools.

The lack of the realization of the dynamic

condition of the industry sometimes causes unjustified criticism. For example, Mr. Young notes the fact that there was neither church nor school in the neighborhood of a large new mill just beginning operations in a rural section of North Carolina. Two years later the corporation had built a school to accommodate four hundred children at a cost of more than $6000, and had aided in the building of four churches. A school was maintained here eight months in the year, half of that time entirely at the expense of the mill, which also contributed to the salaries of the ministers. A hall for lectures and for services of denominations without a church building was also constructed.

This is not at all an isolated case. Other mills have done quite as much. There is scarcely a mill settlement in the state which does not enjoy much greater educational opportunities than the country districts, if only the children could be forced to attend. Some managers parade this support of schools and churches as philanthropy; others say it is a plain matter of business; that the mill which offers the greatest advantages will get more desirable operatives.

The work done by these schools varies. In some the children are well taught and well trained. Where the schools are a part of the city system, the superintendent may give an undue proportion of his time to them on account of the difficulties presented. As the great majority are outside the jurisdiction of the town, there is too often a lack of wise direction. Sometimes the teachers are needy relatives or friends of the mill officers and lack sympathy with the conditions. Occasionally they are of a stern, old-fashioned type, who might give valuable discipline if the children would attend. Usually they succeed only in driving the children away, as parents can hardly keep them in attendance against their will.

The mill child seldom finds school such a welcome relief from monotony as does his country cousin. The instruction seems to him not vital, to touch his life at too few points. There is more fun upon the streets or in the mill than in spending hours in a stuffy schoolroom. He has no traditions to urge him on, and the individuals he admires most may have had little or no schooling.

The night schools established at a few mills either by the management or by individuals have done very little. The operatives are tired after the long day, and there is neither the economic pressure nor the thirst for knowledge which makes such an institution successful in the city. Gradually the attendance, which may have been satisfactory at the opening, lessens until the effort is abandoned.

The churches are, next to the mill itself, the chief centers of community life. The largest in membership are the Methodist and the Baptist. The Presbyterians and the Lutherans have organizations at some mills. The Episcopal church has never had a hold upon the rural population of the middle and western sections of the state, and prejudice against it has been assiduously cultivated. The number of Roman Catholics is negligible.

The power of the church is perhaps greatest in those communities where a large proportion of the operatives is fresh from the country. Often the manager may act as superintendent of the Sunday school, and use his powerful influence to aid the organization. At some mills the corporation itself acts as collecting agent and deducts from the wages the subscriptions which have been made for the support of the work. As a result of this policy, the ministers, who are often men of ability, receive their salaries promptly.

The idea of the institutional church has gained no ground. The church authorities are conservative. The methods used in the country have not been changed to meet the new conditions. Two sermons on Sunday, a weekly prayer meeting, and the Sunday school, are universal. Perhaps there is a missionary society among the women or a "Parsonage Aid Society," and some organization of the young people. These, however, meet with opposition among some of the older members who hold that no organization within the church itself is justified. Some of the Sunday schools have small libraries. The books are usually bought in bulk, however, and are more distinguished for ethical and doctrinal soundness than for literary value.

Old-fashioned orthodox sermons are the rule. The terrors of a literal burning hell, the joys of the righteous hereafter, are expounded with fervor. The emphasis is laid upon the life to come, and upon renunciation of the world, rather than upon a broader, fuller life upon the earth. One minister in charge of a cotton mill church in a burst of impatience exclaimed to me that the mill managers did not wish the thinking powers of their operatives developed, but did wish them to be very religious. This statement is not entirely justified, but undoubtedly the value of religion as an aid to discipline is fully recognized.

Frequent "revivals" are held by the Methodists and Baptists. The churches are filled every night for a week or more, and the services often last until a very late hour. A strange mixture of methods prevails. The "mourner's bench" at which those "convicted of sin" may kneel, and the invitation to shake the hand of the minister as a token of conversion, are both used. The Moody and Sankey hymns, and the old tunes full of haunting minor chords which have done duty at camp-meetings for a century, are heard. Members kneel beside their young friends or move about exhorting them to "come to the altar." The air is electric with emotion, and the old-fashioned type of "shouting Methodist" is not yet extinct.

The pastors of these churches are earnest men, who work faithfully for their charges; but the task is discouraging. Pastoral visiting is unsatisfactory, as often the whole family is never together except on Sunday, and the mother is busy when the call is made. If one family receives a disproportionate share of pastoral attention, the others are jealous. Infinite tact is required, and many ministers avoid so far as possible the care of factory churches.

In spite of all the efforts, the testimony is universal that the churches are losing their hold upon the mill population. The migratory families neglect to bring letters of dismissal from their former churches, and gradually lose their interest in church work. With the increased incidental opportunities for association with their fellows the church services are no longer so important from a social standpoint. In the country, the monthly or semi-monthly services afforded an excellent opportunity to meet acquaintances and friends, who were seldom seen elsewhere. Then, too, the workers are tired on Sunday, and the day is more and more devoted to rest and recreation.

Other agencies may be dismissed with a few words. At a few mills lecture courses are maintained, largely at the expense of the mill management. Theatrical amusements are under the ban. There are few concerts, except, perhaps, those given by the Sunday school. A very few mills have reading rooms which may serve also as clubhouses, but generally the management considers the work of furnishing amusement no part of its duty.

More and more, however, the mills are encouraging care in the surroundings of the tenements. Prizes are sometimes offered for the best-kept lawn and the most attractive flower or vegetable gardens. But vegetation does not thrive upon a sun-baked hillside, when there is no water system, and often there are few competitors. More attention is being paid to the surroundings of the mill itself, however, and some are very attractive.

There is little to gratify æsthetic cravings around the mills generally. There is little pretense of architectural adornment of the mill itself. It is frankly utilitarian, an attempt to secure the maximum of space, light, and convenience at the minimum cost. The tenements are built with the same end in view, and all are staringly new, though vines may render some of the houses more pleasing, and more comfortable as well. Taken as a whole, however, the general impression given by the factory villages is usually one of monotonous ugliness.

In all the social organizations the influence of the managers is apparent. The people have not learned the social results arising from co-operation and organization. Upon the farms the families were necessarily largely self-sufficient. Few forms of neighborhood activity were possible, and time is required for the realization of the increased satisfaction which may arise from collective action.

To prescribe remedies for the bareness of the lives of the mill population is difficult. The settlement idea is not the solution, even if it were practicable. Any gains would be more than offset by the weakening of the sturdiness and independence which would necessarily follow. The operatives are not willing to place themselves in the attitude of expecting and receiving unearned and gratuitous favors.

The institutional church, wisely directed, in

which they might feel a proprietary interest, would have its influence. A change in educational policy, which would fit the instruction to the needs of the learners, would do more, particularly if accompanied by compulsory attendance. Social secretaries, if those sufficiently tactful could be found, might do much for the girls, who need wise direction.

Meanwhile the arousing of ambition to live a broader, fuller life; the substitution of healthy discontent with that part of the environment which is capable of improvement, for stolidity or unhealthy dissatisfaction with all the surroundings, is the problem of the next twenty years.

# CHAPTER X

### THE DEVELOPMENT OF A CLASS CONSCIOUSNESS

FROM the earliest settlement, North Carolina has been marked by a decided individualism and independence of action. The attitude of the inhabitants toward the Proprietors and toward the king as well, was one of neglect and almost of contempt. A regulation not in accordance with their ideas of justice was ignored, governors were driven out, and yet the community as a whole was neither lawless nor turbulent.

Their ideal of government was the theory afterward stated by Jefferson, that the best government is that which governs least. The people have always been jealous of their liberties. Real or fancied oppression would cause an outburst. The citizens were ready to de-

## DEVELOPMENT OF CLASS CONSCIOUSNESS 183

clare their independence of Great Britain before 1776, regardless of consequences, and when the stand was once taken, they could not be moved.

The various elements of the population did not coalesce and, for that matter, have not done so yet. Every political convention sees a renewal of the old contest. Sections, counties, neighborhoods, all stand for something definite, if that be nothing more than an opposition to change. The state has never been a unit except when some great idea fused the people for an instant.

The first constitution was not a democratic instrument. Voting and office holding were confined to the landowning and the taxpaying classes. With them for a brief period the Whig party was influential, but the Democrats, advised by Stephen A. Douglas, made stirring campaigns upon the issue of free suffrage and won just before the Civil War.

But the influence of the Whig leaders was strong, even after their political power had departed, and when *Demos* came into control he was conservative. He was devoted to the Union, to his state, to his church, and to his political party. He was unambitious, not easily moved by adverse circumstances, accepting the apparent inevitable with equanimity, almost with resignation.[1]

The people were opposed to secession, and the state did not leave the Union until forced out; and then held on doggedly, persistently to the end. No hardship, no deprivation, no sacrifice, was too great, yet attempts of the Confederate government to override the rights of the state were resented with spirit and effect.[2]

After the war, the small farmer naturally called himself a Conservative, and afterward a Democrat. This did not mean so much the acceptance of a body of doctrine as the declaration of his belief in home rule and the dominance

[1] For an illuminating study of the psychology of the North Carolina farmer, see Page, "The Rebuilding of Old Commonwealths" (1902).

[2] "History of North Carolina Regiments," Vol. V.

of the fittest. To this day some of the "Old Line Whigs" dislike to call themselves Democrats.

Yet the independence of thought is shown by the presence of a larger white Republican vote than is cast in any other Southern state. Men have in the past braved ostracism for opinions to which they had somehow, perhaps with infinite difficulty, come. Naturally as with a conservative people, party labels have been inherited, but individuals break away. Slowly but surely the policy of the state has broadened. No backward steps have been taken. Every advance has been kept. If it has not moved so rapidly as others, it has made fewer dangerous experiments and fewer mistakes. While less legislation has been placed upon the statute books, less has been repealed.

The shrinkage of agricultural values twenty years ago worked great hardships, which have been mentioned. To the farmer there seemed something wrong, and the succeeding Populistic movement appealed to him.

Apparently governmental policies had favored certain classes, and he, too, joined the corn growers of Kansas in the demand for relief. But in all of this there was a certain hesitation. He was not sure of his ground, though he favored the Sub-Treasury scheme, in accordance with which the government should issue negotiable certificates based upon his products, deposited in government warehouses. It was done for silver, why not for cotton?

Yet this was not a conversion to Philosophic Socialism. He was simply demanding a "square deal." He captured the Democratic party, but his leaders committed him to the Populist party, and his loyalty constrained him to follow them; but not so far but that he could and did retrace his steps. The doctrine of free silver met with acquiescence rather than with enthusiasm, and there were many hard-headed, poor farmers who refused to become converted to the fascinating doctrine.

His difficulty was his inability to think

consistently of himself as belonging to a class with distinct interests. He felt himself a citizen, equal to any one, and bowed no more to the tyranny of his own class than he did to the tyranny of aristocracy. The farmer legislator was no more inclined to pass special legislation for himself than for a corporation. A certain intention of doing rough justice was inborn. Fairness rather than the desire for personal advantage was his dominant trait, next to his dislike of change.

All this is not inconsistent with what has been said of the influence of individuals. They possessed the influence because they did not demand it, because a reputation for sanity and straightforwardness had been gained. The illiterate or unread man trusted them, not because he felt them his superiors, not because of any claims of descent, or of wealth, but because of his confidence in their wisdom, and in their character as citizens. No dishonest man has ever held political influence long in this state.

When the farmer, for any of the reasons mentioned elsewhere, comes to dwell in the mill village, the difficulties of adjustment occupy him for a time. He cannot live his old life, and his place has not yet been found. If he works himself, steady employment, day after day, rain or shine, contrasts strongly with the irregular employment on the farm. If he lives upon the earnings of his children, the unaccustomed leisure is no less strange.

The substitution of a wage economy for a products economy is strange also. On the farm the family produced what it ate. At the mill it purchases with money what is consumed. The amount received, measured in money, is so much larger than the former income that an impression of prosperity is substituted for one of scarcity. This is true even where the residual income is no larger than it was upon the farm.

The operative farmer may begin to question some of the labels which he has always borne. He begins to see more clearly the intimate connection of governmental policies with economic interests. One mill manager said, "My men are nearly all Republicans, if they only knew it." Party ties, though not broken, hang more loosely. Independent voting becomes more common.

Socialism makes little appeal to him. He listens to an occasional missionary, but the arguments make little impression. For one reason he is too close to the soil; for another he has had in the state as a whole no real contest with the employer in which capital was arrayed directly against labor. In few counties has the Socialist party even a skeleton organization. In one county 84 votes were cast in 1902. In the whole state, Debs received only 124 votes in 1904, and few of these came from the cotton mills.

As might be expected, a class not yet conscious of itself affords sterile ground for labor organizations, which must be based upon individual subordination. The workers have always acted as individuals and have not learned the power of collective effort, nor have they felt the compelling necessity. The right of a man to live his own life, subject of course to the moral law, but unhampered by any other restrictions, seems to him obvious and natural.

In 1897 the Arkwright Club of Boston predicted the ruin of the New England industry unless the advantages of hours and wages in favor of the South were lessened. In 1899 Mr. George Gunton, after a trip through the South, advocated the raising of a fund by Northern manufacturers to be spent in unionizing the Southern operatives, and there is reason to believe that some money was contributed for this purpose.

Organizers from the North were sent through the South, and local organizers were appointed. Unfortunately for the cause, many of these local appointees were men of bad character, unfrocked clergymen and the like, who did not command the respect of the people. The protection of the American Federation of

DEVELOPMENT OF CLASS CONSCIOUSNESS 191

Labor was promised to the unions, and glowing prospects of shorter hours, increased pay, and greater privileges were pictured. Unions were organized in a number of mill settlements, though in no case did the organization include the whole labor force. The operatives were generally content, particularly as wages had been increased about 10 per cent. during the twelve months preceding.

The managers immediately took counsel and agreed to act promptly. Heretofore they had managed their enterprises without dictation, and now felt that a crisis was at hand. The time was propitious for them. The market was overstocked, and many mills were either making very small profits or were operating simply to keep their force together, thankful if they could "swap dollars."

The first contest came in Greensboro. There the Northern owner of the Proximity Mill, who said that he had come South to avoid labor troubles and to be free from the tyranny of the textile organization, on learning of the existence of a union among his operatives, promptly closed his mill. The union was not prepared for such tactics, and was soon dissolved.

An employee of the Erwin Mills at Durham left the mill on business for the union, though permission had been refused by an overseer. He was promptly discharged on his return, and, moved by a sudden impulse, the men went on strike. When an appeal was made for food for the strikers, a week or two later, Mr. Erwin announced that his contest was not with the men, but with unwarranted interference with his business, and authorized the merchants to issue supplies to all his employees, including the members of the union. No man in the state has done more for the welfare of his operatives than he, and this fact, coupled with his somewhat quixotic action, soon caused the disbanding of the union.

The strongest conflict occurred in Alamance County, where there were nearly twenty-five small mills, owned chiefly by the Holt family and its connections. The discharge of an unpopular overseer was peremptorily demanded in the mill of the T. M. Holt Mfg. Co. The refusal resulted in a strike, and the operatives in another mill owned by the same company struck in sympathy, and further strikes were threatened.

After consultation the representatives of seventeen mills posted the following notice during the latter part of September, 1900: —

*Whereas*, recent developments have shown that this mill cannot be operated with that harmony between the owners and the operatives thereof, which is essential to success, and to the interests of all concerned, so long as the operatives are subject to interference by outside parties, this is to give notice that on and after the 15th day of October, 1900, this mill will not employ any operatives who belong to a labor union, but will be run by non-union labor only. All operatives who object to the above and will not withdraw from labor unions will please consider this as notice and vacate any house and premises, belonging to us, which they now occupy, on or before the 15th day of October, 1900.

The union at one mill weakened immediately, but sixteen establishments were closed, and the manufacturers thus avoided piling up unsalable goods. Few of these mills had charged rents for tenements, and the strikers had been living in them rent free as usual. The order of eviction was not carried out strictly, but many operatives left, some to go back to the farms, some to other mills, chiefly in Georgia.

The national organization took charge of the strike, and one Thomas was sent to take control. The mill owners absolutely refused to treat with him, however. Some aid was sent by the national organization, and more by the Southern unions; but the demand was much greater than the supply. There were few houses for the strikers after their eviction, and no money to pay rent. No disorder occurred, and there was no call for the exercise of police authority.

About the 15th of November desertions from the union forces were so numerous that the organization collapsed, and the workers sought to return as individuals. Many nonunion operatives had already been employed, however, and these were not discharged. The vacant places were given to the least obnoxious of the strikers.

Meanwhile a threatened strike at Fayetteville had been crushed by similar tactics. At many other mills, meanwhile, employers discharged union men. Where this was not done, the unions were awed by the continued success of the manufacturers, and made no demands. A number of unions continued to exist; but the number is gradually lessening, and those still existing are weak. Some manufacturers will employ no union men. Others, who are confident of their strength, are indifferent. In a few mills, some of the operatives are secretly organized.

Loyalty to the cause is not yet strongly implanted. In fact, it is hardly realized that there is a cause. To the average operative the gains from unionism are not balanced by the sacrifice of individual initiative, the right to independence of action. He is not yet ready to put the union before everything else. He does not feel that the spinner in Fall River or the weaver in Lowell are closer to him than the people of his own section even though they pursue different occupations. In 1903 the United Textile Workers practically threw overboard the Southern unions, chiefly on account of failure to pay assessments.

The opposition of the manufacturers has the same foundation as their opposition to restrictive legislation. They do not consider their business a public matter, and resent interference. They believe that the organization of their mills was an attempt to cripple them, on the part of their competitors, and that agitation for shorter hours, a national labor law, etc., arises from the same source, and not from any real interest in the welfare of their operatives.

They have heard of the excesses of the unions in other sections, where strikes lasting

## DEVELOPMENT OF CLASS CONSCIOUSNESS 197

for months are not uncommon. Their operatives have been generally content and have worked without friction. Naturally they wish to preserve this condition as long as possible. For this reason many opposed any beginning of legislative interference. Few object to the law passed by the legislature of 1903, but they fear further encroachments upon their liberty of action.

The position is thus set forth by a prominent manufacturer: "The entering wedge has been made, and now we look for each successive meeting of our honorable law-makers, and guardians of the people's rights and liberties, to become more radical, until in the course of a very short time we shall see the 'walking delegates' in full force and control. Equality of opportunity is the sole distinguishing feature of American civilization, yet we see a supposed conservative body of representative North Carolinians, unknowingly abridging this principle of liberty and laying the mud-sills upon which will germinate unions and all of the attendant evils connected with the same, which are becoming dangerous not only to their original purpose but to our very government itself. Our competitors of the East are now very much entangled in the web of labor laws and desire that we be caught in the same net."

Some manufacturers, ignorant of industrial history, believe that by united effort the unions can be kept out permanently. The inconsistency of their position does not occur to them. Others realize that the organization of the mills is a matter of time, that a class consciousness must develop more rapidly in the future; but all are resolved to postpone the day as long as possible.

Meanwhile there are signs of restlessness among the operatives, slight to be sure, but existent nevertheless. No particular grievances have yet been formulated. The questions of shorter hours and decidedly larger wages have not yet become demands. In fact, they are hardly yet seen as possibilities. But there are signs of incredulity when the employer speaks of low profits or difficult sales. They wonder at the success of the manufacturer as yet, but envy may follow, and then questioning.

When the operatives know no other life than that of the mill village, when the connection with the soil is broken, leaders may arise who will preach the "war of the classes." Much preaching will be necessary, but the dormant class consciousness is already stirring. Already the operatives are beginning to think of themselves as a peculiar people. Many years, however, will be required to produce sharp lines of distinction. There is too much inborn and inbred democracy for the final breach to appear in the immediate future.

## CHAPTER XI

#### THE RELATIONS OF EMPLOYER AND EMPLOYED

THE relations of employer and employed in the small mills of the South, and of North Carolina particularly, differ from similar relations in the older manufacturing sections, both in spirit and in detail. Manufacturing has developed so rapidly in an agricultural society that the relation is still one between individuals, and not between a corporation and a class.

There is, as yet, no self-created aristocracy of mill owners, such as developed in England or New England with the growth of the cotton industry. The second generation seldom has charge of mills, and still more rarely the third. The active manager of the mill business, himself usually a stockholder, is not dependent upon non-resident stockholders. In the small mills the manager, either the secretary or the president, often knows every operative both by sight and by name. He may know any kinship between different families, and may even know the quarrels and the love affairs.

Often the manager and the operatives have many experiences and traditions in common, and share many of the same ideas. The stages in the progress of a mill manager are often from a farm to a country store, from the country store to a village, and then to the control of a mill. Possibly as a boy he played with individual operatives or with their parents in the country, and knows them for two generations. At least he knows the general neighborhood from which they came, and can understand their point of view. I have heard an operative call a mill manager by his first name, without a suspicion of impertinence. There is, too, a certain clannishness among the Southern people, which sometimes has practical results. It is related of one successful manager of several mills, that he will always manage to find work for a family from the county in which he was born and began his business career. Then, too, there is the broader feeling that they are all of the same people — Southerners; sharers in sympathy, history, and traditions. The general result is kindliness on the side of the employer, and loyalty on the part of the employed.

Almost invariably in the smaller factory communities, and often in the larger, the house of the mill manager is near the mill and its tenements. The manager and operatives attend the same church and Sunday school. The manager's wife may be active in church work, and often visits every sick woman or child in the community. The manager may take an active part in the support and management of the baseball team, practically the only form of athletics commonly found. When an operative is in trouble, he instinctively turns to the manager for advice, and usually receives aid as well.

With the knowledge of the circumstances

and the residences of every family comes inquiry if operatives are missing from their places. Miss Van Vorst totally misunderstood the purpose of such inquiry in the sentimental and imaginative account of her experiences.[1] It is not prompted by harshness but by the desire to secure the knowledge necessary for successful operation. When sickness comes, supplies are furnished from the store, if one is connected with the mill, or accounts are guaranteed at a neighboring store; fuel is provided; a doctor is sent, and funeral expenses are guaranteed. The mill is protected to some extent by the wages kept in arrears, but in any considerable illness this is soon swallowed up, and only the innate honesty of the operatives prevents loss.[2] Sometimes the confidence is abused,

[1] "The Woman Who Toils" (1903).
[2] Wages are paid weekly or fortnightly, but generally for the previous period; *i.e.* an operative beginning work receives no pay until two periods, or else a fortnight and a week, have passed. Then wages are paid regularly. Notice of leaving is required, and if all the accounts due the mill have been settled, the arrears are paid at once.

but oftener the workers return to the mill and pay the accounts.¹

Where the mill owner or manager will take the trouble he can usually influence the votes of his operatives on any question not immediately connected with the partisan politics, and sometimes even then. In 1896, as the result of a bitter local contest, the operatives of a large mill voted the Republican ticket solidly. Such a departure from the usual course is rare, but on local questions the operatives vote with their employers. Many towns have laws prohibiting the sale of intoxicants through the powerful influence of the mill management, which demands them as aids to the regularity of operatives. In short, the relationship is more or less feudal. The manager is the strong man, who has forced himself to the front, and proves his

---

[1] One superintendent informed the writer (1904) that within two years, during epidemics of grippe, he had guaranteed accounts amounting to nearly $3000, of which he had been called upon to pay less than $5.

fitness for leadership by his continued success. On the other hand, it is not fear which influences the operatives. They follow leaders because they respect them, because it seems the natural course. Possibly it is inherent in Southern character to look to men rather than to an abstract consideration of relationship.[1] The maintenance of such influence grows more difficult with the increase of population in the towns, unless the mill is upon the outskirts and has its own community life.

There are few mills where an operative having a grievance against an overseer cannot take it directly to headquarters and secure a hearing. The employer does not screen himself behind office boys, but will deal with the operative man and man and not refuse him the opportunity to present his side of the case. Overseers are often overruled and discipline is benefited rather than injured by such action. In spite of their deference to the man-

[1] For further discussion, see Chs. VI and X.

agement, the operatives are not servile. The expression often heard about a mill, "'pore,' but proud," is wonderfully descriptive. The pride is foolish oftentimes, but they will give up a position rather than submit to a real or fancied injustice. The confession of a mistake and its correction by the management will be less harmful to discipline than sustaining a manifestly unjust act of a foreman. Of course, injustice may be done. The word of a superintendent or an overseer naturally carries more weight than that of a less important employee; but an overseer no matter how great his technical qualifications cannot hold his position very long when disliked by any considerable portion of the operatives. The personal element is so important in the Southern mills that a good manager of "help" is more valued than a man having greater technical knowledge without such tact.

For this reason the importation of New England overseers has not been particularly successful. Apparently these are accustomed

## RELATIONS OF EMPLOYER AND EMPLOYED

to look upon the operatives, during working hours at least, as a part of the mill equipment and to neglect personal peculiarities. They forget that they are not dealing with crystallized mill traditions developed through a hundred years, but with individuals fresh from rural independence, engaged in a new industry. These individuals, while loyal and tractable, are at the same time restive under control. The case of the regular and the volunteer regiment presents some points of similarity, but the mill superintendent cannot hold his operatives against their will as the army officer is able to do. Another position is too easy to secure.

Even if these operatives were only tenant farmers, large liberty of action was allowed, in details at least, and they cannot be treated as a mass when they come to the mill. A part of the difficulty experienced by the New Englander may be owing to the feeling of aloofness and to the sectional prejudice of the Southerner, particularly of the uneducated Southerner — an aloofness which would require a volume to explain. However, certain individuals among the imported overseers have been remarkably successful, and this would seem to show that the difficulty does not lie in sectional prejudice alone. No better explanation than that of the attitude and personality of the foreigner generally has been offered.

There is little definite jealousy of the greater wealth of the operators. One may hear around the village many complaints of "hard times," or of the smallness of the wages. The prices paid at the company store may be criticised, though, in fact, they are so much lower than those paid when buying "on time" in the country, that this complaint is not heard so often as might be expected. The company store in North Carolina usually meets the competition of independent establishments and often attracts the custom of outsiders. Years ago operatives at many mills were paid in checks, good at the store, but redeemed in cash at unfrequent intervals. Prices were frequently exorbitant, but economic forces have brought great changes except in a few remote localities. The company store is no longer universal and often exists principally for the convenience of the operatives. A cotton-mill superintendent of wide experience states the opinion that the prices in the company stores generally range from 5 to 10 per cent. lower than in the independent establishments,[1] and personal observation seems to confirm this view. This is possible, since they cater to a definite trade, can estimate closely the volume of business, and sell only for cash or its equivalent.

The justice of the local distribution of wealth is seldom questioned. They have seen the capitalist apparently create wealth unknown before. They saw the better market for wood, cotton, and farm produce follow the establishment of the mill. Their own wages

[1] Interview, 1904. See also pamphlet, "Do Not Grind the Seed Corn" (National Child Labor Com., 1905).

P

buy comforts and luxuries unknown on the farms, and these outweigh to them the advantages of country life. It is simply a phase of the problem which vexes the social student in the city, who laments the fact that people will not stay upon the farms, but prefer to live in crowded tenements. As the members of the family grow older, the income increases and a rising standard gives more satisfaction than a higher stationary one. Put into the succinct phrase of a native, "It's not doing well, that makes people contented, it's doing better." More than all else, perhaps, these people and their ancestors have held land, and a land-holding population is not inclined toward socialism. As these traditions of a rural past are lost, in the development of an operative class, two or three generations from the farm, and wants increase faster than the means of supplying them, changes will come.

The attitude of the mill owners and managers toward the social welfare of the operatives is by no means uniform. At some mills much is done; at others, nothing except perhaps a subscription from the mill treasury to aid a struggling church. Existing agencies for social betterment are described and discussed in another place. What has been done is usually the personal expression of the man in charge who controls the directors while he is able to pay dividends. He has not been coerced by public opinion, nor by the demand of the operatives themselves, nor by a sense of the collective responsibility of the corporation. The public as yet recognizes only two factors in industrial life, — the state and the individual. It has not become conscious of the corporation, nor fixed its responsibilities. So far the corporation is only the men controlling it, and they are judged in a personal rather than in a corporate capacity.

The corporation has not recognized its quasi-public character. The stockholders and directors still consider their relation to the operatives that of individuals. The duty of an individual, they consider, is to pay the market rate of wages, and in addition to give personal kindness in misfortune and protection in trouble. Their conception of the duty of a corporation is similar. The difference between the condition of a few individuals, with many outside interests, working for another individual, and that of a large body of operatives of both sexes and all ages in a community of their own, working, year after year, for a corporation, around which all their interests center, has not been realized. There are striking exceptions, particularly among the larger mills, but the corporation idea is too new for any considerable growth of other sentiment.

If the managers of the mills have not felt the obligation to provide elaborate agencies for social improvement, the operatives, on the other hand, have not definitely demanded them. They, too, are individualists, even though they have become gregarious. Many are ignorant of such things, and they are suspicious of anything which smacks of patronage. The Social Settlement, with the sense of superiority thinly covered with a veneer of fellow-feeling, would find little support among them. They have not yet become accustomed to expect that which does not come as a direct tangible result of their own labor. Social work in a factory community is difficult, requiring infinite tact. The tentative efforts put forth in some neighborhoods have often resulted in failure, to the discouragement of the experimenters. This failure has not been caused so much by ingratitude as by independence. The operatives recognize no reason why these things should be done. Their employers are only men in better circumstances than themselves, and not a class raised above them from whom they must gratefully accept tokens of good will. The manager must come man to man if he wishes to influence them.

These personal relations cannot endure. The isolation of the cotton-mill communities is already breaking up in a few localities. Heretofore the cotton mill often has been the only industrial establishment in the town or village; but with the rapid development of other manufacturing enterprises, the textile operative becomes only one of a number of industrial workers. Up to this time their association has been instinctive rather than rational. With the development of a class consciousness will come a weakening of one side of that relation to their employers which, for want of a better word, we have called feudal. On the other hand, the lessening of pride and independence among the weaker, and the development of dependents and paupers, is to be feared.

The previous futile attempts to organize the mill workers are described in another place. With the development of a class consciousness, organization follows naturally. The employers are already organizing. The profits are smaller than ten years ago. In some lines there is overproduction; labor is not so abundant, though the reserve is still large; the margin between the cost of the raw material and the price of the finished product is not so large. Each manufacturer can no longer be a law unto himself. Associations and understandings become common. These range from informal agreements among the manufacturers of a town, regarding employment of operatives or the purchase of cotton, to larger organization attempting to control prices and production.

The increasing complexity of his problems is working changes in the manufacturer. Individuals of a type before almost unknown in the South, though common in industrial societies generally, appear here and there. They are cold, shrewd, farsighted. Sentiment in them does not interfere with the strict working of the principle of self-advantage. Sometimes he is a man whose family was long prominent politically or socially. The stress of circumstances following the Civil War possibly deprived him of educational opportunities, and the struggle for existence has embittered him. The reaction from the old ideas is almost a revulsion. Scrupulous exactness in financial matters has been substituted for carelessness; the desire for wealth, for political ambition; an eager activity, for ease and dignity. Perhaps he has studied the position of the entrepreneur in other sections and discards former relations as Southern provincialism. Sometimes this manager has come up from generations of poverty and hardships. He has conquered his obstacles and intends to make the most of his success. With these must be mentioned a few Northerners, who have come to the state to escape the restrictions in the North.

This type of employer is not yet common and is not likely to become the rule. The personal element in Southern life is too strong. Meanwhile, however, his influence upon the managers of the older type must be felt. Another type, a compromise, will arise, which will deal with a new class of employees. In the inevitable contest for a larger share of the product, the operatives will, for a time, lose more in personal kindliness than they gain in wages or hours of labor. The organization of the employers will be perfected first. Slowly the operatives will sink their individualism and independence in an organization of which the benefits are not immediately apparent. When the lesson is learned, the unions will be powerful. These men are physically fearless, are native to the soil, are capable of sacrifice for an idea, and there is always the land to which they may return if beaten.[1]

Undoubtedly many manufacturers do not realize that the older type of operative, or of father of a family of operatives, is passing, never to return. The changes in the attitude of the operatives are coming slowly, and are by no means uniform in different sections, or even in the same section. At one mill

[1] See Ch. VIII, p. 159.

the old patriarchal relation continues; at another the labor force is constantly changing, and the only bond is the "nexus of cash payments."

## CHAPTER XII

### THE CHILD IN THE MILL

THE impression that the success of the Southern mills has been built wholly upon the labor of children is widespread. Real and fancied descriptions have emphasized this view. The labor agitator, the professional reformer, the yellow journalist, have joined in telling of "childhood blotted out to add a few more dollars to the dividends of aristocratic stockholders of hell's mills of the South." Certain phases have been seized as themes for fiction.

The defenders of Southern conditions have resorted to denials, to heated denunciations of meddlers, to rose-colored descriptions, and to countercharges. Their answers have been little read except locally and have done little to modify the prevailing idea. A few persons have endeavored to describe conditions faithfully and dispassionately; but they, too, have made little impression, perhaps because they have not sought to analyze the underlying causes.

Now child labor is first of all an economic phenomenon and has existed through all stages of civilization. The basis is economy of muscle and intelligence in applying to a given task no more of either than is necessary for satisfactory accomplishment. In Egypt and India, at the earliest periods of which we have any account, children had their tasks. They have had them in all agricultural societies and have them to-day.

Industrialism does not create the phenomenon, but concentrates it, and changes the form, regularity, and intensity. A larger proportion of children is employed in an agricultural than in a manufacturing society. On farms nearly all children old enough to be of service work all day during a part of the year, and a part of the day during the whole year. In 1880, before the growth of manufacturing in North Carolina, 55.9 per cent. of the males between 10 and 15 were reported as being engaged in gainful occupations, compared with 55.1 per cent. in 1900, and 43.1 per cent. of the total was still engaged in agriculture in this year. But even these figures, large as they are, do not represent the whole truth. Undoubtedly many farmers did not report the names of all their children who performed services upon the farm.

This has always seemed the natural state of things in an agricultural society. It is only when the people of a community have ceased to think in terms of agriculture that the discussion of child labor as an ethical and social problem begins. Only then are social considerations contrasted with the apparent economic advantages, and we find the theory advanced that no work whatever should be required of children.

It is in an agricultural society suddenly

engaged in manufacturing that we must study the employment of the child. The problem is not new. We shall find that the employment of children in the North Carolina mills follows well-defined laws observed elsewhere, modified, however, by certain social and sectional idiosyncrasies, which must be considered.

The number of children engaged in manufacturing is difficult to ascertain, even though their employment is confined chiefly to two industries, — textiles and tobacco. The latter and smaller industry is irregular in operation, and we shall confine the discussion to the manufacture of cotton, including, however, a few small woolen mills.

The first difficulty lies in the definition of child. The United States Census fixes the line between adults and children at 16, and its tables are made upon this basis. For a time the North Carolina Bureau of Labor Statistics fixed the line at 14, but since the agitation for the present act prohibiting the employment of those under 12 began, only those below that age have been so counted.

In 1900 the Census reported 7129 children below 16 engaged in cotton manufacturing out of a total of 30,273 operatives, or a little more than $23\frac{1}{2}$ per cent. The State Bureau of Labor Statistics for the same year found in the entire textile industry 7598 children below 14, of a total of 38,637 operatives, or about $19\frac{3}{4}$ per cent. of the total. The next year 7996 below 14 were found in a total force of 45,044, or about $17\frac{3}{4}$ per cent.

In 1901 the manufacturers, to prevent the passage of a labor law, formally agreed to employ no children under twelve (unless the support of a widowed mother or a physically disabled father), and the statistics for 1902 were gathered upon that basis.

In that year, 178 mills representing 1,209,819 spindles reported 929 children under 12, an average of a little more than 5 to the mill. Ninety-eight mills with 533,612 spindles failed to report this item. If the same proportion existed in these mills, the total in all the mills of the state would be 1339. It is natural to suppose, however, that the omission was intentional and the obvious reason would appear to be the disinclination to give evidence against themselves. Miss Sewall found in 1903 that the number of children under 12 in a limited number of selected establishments was about 18 per cent. of those under 16, which proportion would make the number for the whole state about 2000.[1]

The figures of the mills reporting to the state bureau are, moreover, not above suspicion. No vital statistics are kept by the state nor by the municipalities. Where the managers honestly attempted to prevent the employment of very young children, the word of the parent was necessarily the chief reliance for information. Always in industrial history such statements have been influenced by greed or apparent necessity.

[1] U. S. Bureau of Labor Bulletin, No. 52 (May, 1904).

The figures given in the earlier reports, *i.e.* the number of those under 14, are perhaps approximately correct, as in them there was no reason for misstatement. The comparison of figures from all available sources apparently shows that about 25 per cent. of the total number employed are under 16 years of age. This proportion, while large, is much smaller than has been indicated by irresponsible writers.

Observation regardless of statistics shows the extensive employment of children. A visit to certain departments of a mill shows few adults; or one may see the children entering or leaving the mills as work begins or ends. In fact, their numbers often seem proportionally greater than they actually are from the concentration of their employment in a few departments.[1]

Further, many of these children began work at a very early age. Of 98 children observed by Miss Sewall, 79 began work before the age

[1] See Ch. VII for description of work.

of 12, and 37 began before the age of 10. Only 11 under 12 were found employed at the time of the investigation, however. These figures are not broad enough for a generalization, and some of the establishments visited are among the worst in the state.

As was indicated in a previous chapter, these children are employed almost exclusively in the spinning rooms. Few in other departments are below 16. The work has been described. The spinners, chiefly girls, twist together the broken threads, and the boys replace the full bobbins with empty ones.

From the standpoint of muscular exertion, the work is not difficult. Little physical strength is required, and the work is not continuous. The position while working is constrained, however, and backs grow tired from stooping before the long day draws to a close. When not employed the spinners may sit, and the doffers lounge about the mill, or even play in the yard. The air is fresh, though sometimes filled with tiny particles of lint, depending somewhat upon the product and the quality of the cotton.

The chief demand is constant watchfulness. Every thread must be mended as soon as it is broken, and a spinner has from 200 to 800 spindles to watch. No great demand is made upon the mental faculties, and the work is monotonous. The noise of the machinery also must have its effect upon the nerves and indirectly upon bodily well-being.

Speaking broadly, the physical effect of the work is undoubtedly bad, though not all are affected unfavorably. Many of the children employed are sturdy and strong and one may find many men and women of good physique who have worked in the mills from childhood. Apparently neither the confinement nor the noise and tension have affected them unfavorably. Their lungs have not been affected by the lint, and they have vitality enough to overcome the disadvantages of their employment. The pitiful stories told by "two ladies" who became "factory girls" for a few weeks have little real foundation. They sought sensation rather than truth, and found it.[1]

Compared with some other forms of child labor, the conditions in the cotton mill are not unreservedly bad. The work is probably less exhausting than that of a cash boy or girl in a busy department store, where the air is almost invariably worse than in a mill. The newsboys in the street suffer more physically and morally.[2] The children employed in glass factories and coal mines have work that is more injurious in every respect.

But nevertheless the confinement for the long hours, combined with improperly chosen and badly prepared food, is enough to stunt the growth and lessen the vitality of a great number. This is particularly true of night workers. Refreshing sleep during the daylight is difficult amid the noises of the factory village, and sometimes the children who finish their tasks at six in

---

[1] Van Vorst, "The Woman Who Toils" (1903).

[2] "The Street: Its Child Workers," University Settlement Society, New York.

the morning are up again when the day workers return for their dinners at twelve. It is not surprising that children sometimes fall asleep over their tasks, and that the work is poorly done. Yet a majority of the children, perhaps, prefer the night shift on account of the few hours gained for play in the afternoon.

The disadvantages of night work are recognized as many and serious, and the amount is decreasing. One manufacturer writes, "Neither evil (child labor nor illiteracy), nor both together, is half so great as night work for women and children." Another says, "Night work hurts worse morally than it does physically, and every sane man knows what a strain on the system night work is." Unfortunately the departments commonly operated at night are those in which children are extensively employed.

The mental effects are both positive and negative. Probably the monotonous routine introduced so early into their lives has a tendency to hinder mental development, though teachers differ upon this point. Some declare that children who have worked in the mill are more eager to utilize their opportunities and accomplish more than other children. Others find them difficult to teach. Probably the difference depends somewhat upon the length of time spent in the mill before entering school. Certainly the power of initiative is lessened.

The greater injury is the deprivation of educational opportunities. A child who goes into the mill is too often a fixture. If illiterate when he enters, too often he remains illiterate. If he has only the rudiments of an education, the long hours give him little opportunity to extend his knowledge, and instances where a child has forgotten how to read are not unknown.

In the country the children might attend the short sessions of the public schools which were taught when farm work was slack. The mills run continuously, school attendance means the loss of wages, and some ignorant parents see no compensating advantage. Of the 98 children reported by Miss Sewall, 12 had not attended school before entering the mill, 64 had not attended afterward, and 8 had not attended at all. A thriving manufacturing town which has excellent school buildings and schools reported an enrollment in the public schools of 35 per cent. of the population between 6 and 21. The enrollment for the county, excluding the town, was 67 per cent. However, this comparison is not absolutely fair to the town, as it contains several private schools for both white and colored, the enrollment of which was not counted. The percentage in boarding schools was also larger in the town than in the country. But the addition of these students would still leave the proportionate attendance in the town lower than in the rural districts.

The moral effects of the work depend somewhat upon the mill. Generally the manager strives to keep out evil influences. Profane and immoral overseers are seldom tolerated, and immorality among the operatives is not countenanced. One observer says: "The moral atmosphere of a mill settlement is much purer than I have ever seen it in the North. People with bad habits or inclined to lead disorderly lives are not tolerated."[1] Many instances of this policy have come under my own observation. But when all possible has been done, the children see and hear many things which do harm. On the other hand, the discipline of work teaches them lessons of obedience, carefulness, and self-restraint.

Nor does the work entirely destroy the spring and elasticity of childhood. Many hate the work, of course, but the general attitude of the children is not one of rebellion. On the farms the children worked, and they accept their occupation as a matter of course. They take pride in being wage earners and treasure a word of approval from an overseer or superintendent. Their ambition is to be transferred to the looms, where they make larger wages.

[1] Dr. James C. Bayles, in *New York Times*, June 2, 1901.

Many prefer the mill to the school, and will attend only under compulsion. When the child has been contributing to the family support, and is certain of employment at will, such compulsion is with difficulty applied by the parents.

On this subject, perhaps the testimony of the late Dr. Charles B. Spahr is again worth quoting. Speaking of children at work he said, "They went about their work with so much spring and seemed to have so much spirit in it all and after it all, that I was completely nonplussed."[1] Experiences of my own with mischievous doffers confirm the statement.

How do parents justify sending their children to such employment? As mentioned in a previous chapter, the migrants from the farm are made up of five classes: (1) the honest man seeking to better the condition of the family; (2) the incapable or shiftless; (3) the disabled; (4) the lazy; (5) the widow. Each class must be treated separately.

[1] "America's Working People" (1900).

A man of the first class reasons that the rewards of the family labor on the farm are low at best, and always exceedingly variable, since floods, droughts, or other causes may destroy all hope of profits. At the mill, work is permanent and definite wages are paid in cash. Instances of $20 to $30 dollars a week earned by a single family seem unlimited wealth, since he seldom considers the value of provisions consumed upon the farm when comparing his lot with that of the factory family.

His children work with him on the farm, hoeing or plowing during the long, hot summer days; they pick cotton or haul wood when the frost bites the fingers. He thinks that it will be easier for them to work indoors. Then, too, the public school in the country is open barely four months in the year, and it is often two or three miles from his home. In the factory towns the schools are good and the term ranges from six to nine months, since the school fund is almost invariably supplemented by the corporation when necessary to produce this result.

He does not intend to exploit his children. He intends to work himself and, when the family goes to the mill, seeks employment. After forty years or more spent upon the farm, however, his roughened fingers can seldom be trained to do work requiring dexterity, and the number of common laborers required is limited. Perhaps he secures employment as teamster or truckman for the mill, or a position elsewhere in the town is found. In many cases the cotton mill is the only industrial enterprise in the vicinity, and there are few openings which afford steady employment. Meanwhile his older children are earning more than he can command.

Other men around him are not working, and too often the unemployed periods grow longer. Familiarity with the idea of being supported by his children blunts his sensibilities. He salves his conscience, perhaps, by cultivating a garden, chopping the wood, and doing the chores. Sometimes he secures the agency for some patented article, but he seldom possesses the qualities necessary for a successful agent, and this occupation often serves only as his excuse for his lack of steady employment; or else he discovers some ailment to excuse his idleness. The result, of course, is moral deterioration.

Meanwhile he finds that though the family income has increased, expenses have increased also, compared with the farm. Numerous new wants have become a part of the standard of life. Sickness with its attendant expenses may come to a child, causing loss of wages as well. Perhaps the older children marry and their wages are no longer a part of the family income. The education of the others may be postponed from year to year, until sometimes they feel ashamed to enter the primary grades and refuse to attend if the long-delayed opportunity finally comes.

The history of the man of the second class, the incapable or shiftless, is similar, except that he more rarely finds permanent employment, and sooner becomes content to be a drone. He may be a man of the best intentions, obeying the moral law as he understands it, and faithful in religious observances, but his will power is not sufficient to enable him to resist the influence of his surroundings.

If a strong man has difficulty in securing employment, much harder is the lot of those partially disabled by old age, exposure, or disease. The managers often make places for them as sweepers or messengers, of course at low wages. Oftener they busy themselves with trifles outside and gradually cease to do even that work for which they are fit.

The fourth class, the lazy, deliberately living lives of ease from the labor of their children, is small at first but constantly receives accessions from the divisions already named. These men loaf about the stores or the blacksmith shops, discussing politics and gossiping. Some do not work at all, even paying for blacking their shoes from their children's wages. Some have a pretended ailment as an excuse; others frankly say that they worked to bring up their children during their early years, and now they expect the children to support them.

Though some of these men are, in a way, kind fathers, necessarily their interest in their children becomes mercenary. They speak of their children as property. "She was the best spinner I had," is an expression not unknown. In sickness their chief concern seems the loss of wages. Such fathers oppose labor legislation, and oppose also the marriage of their more skilled daughters. Runaway marriages often result, as the country girls coming to the mill after the age of sixteen usually do marry.

The mill managers may declare that they will not have such men about the mill, and may compel them to undertake some work. Often the unwilling workers make themselves so inefficient that they are necessarily discharged. Good operatives are not too plentiful, and if the family is sent away, the vacant places may be filled with some difficulty.

The legislature of 1905 passed a vagrant act aimed at these able-bodied vampire fathers living upon the earnings of minor children, but provided no adequate means for its enforcement. The problem is one of the transition period. The phenomenal growth of the industry cannot continue. Fewer whole families will be brought from the farms, as a mill population develops. The boy trained in the mill accepts the mill work as a matter of course when he becomes a man, and will be able to earn much more than his farmer father suddenly taken out of his environment.

The following quotation from an English source is applicable here: "Kind-hearted people, too, may follow a course of conduct with their own offspring which appears monstrous to a stranger. In certain districts where child labor is a tradition and a custom, the very idea of associating it with inhumanity does not occur to the people."[1]

It is true in North Carolina. Men who would resent the charge that they are cruel or unnatural parents, press their children into the

[1] "Dangerous Trades," New York and London (1902).

mills often against the desires of the authorities. Long before the passage of the act establishing the age limit at twelve, many mills had already voluntarily established that rule, but the pressure against it was steady. A father would declare that the family must have the additional income from the labor of a child, and threaten to move elsewhere if the child were not admitted.

The fifth class, the widows, exists, though undoubtedly the number dependent upon the labor of very young children is not so great as the opponents of restrictive legislation claim. The lot of such a woman is hard. She toils, cooking, sewing, scrubbing, during the day and sometimes may lie awake at night thinking of her child at work. There seems no solution, except to call upon charity if the children are debarred.[1]

The attitude of the employers of child labor is not uniform. Some realize the social re-

[1] Compare Professor F. H. Giddings, Address before National Educational Association, July, 1905.

sponsibility, endeavor to keep the numbers as small as possible, and encourage school attendance. Others consider these matters to belong primarily to the parents. The educational advantages of some employers were limited. They worked themselves as children, are proud of their mills, and really believe a child is better off in a mill than idle upon the street. The employers generally claim that the labor of young children is wasteful and uneconomical, and that they are employed only because of the demand or necessity of the parent, and to keep them out of mischief.

The following extract from a frank letter is the expression of a man who worked his way to the position of manager and part-owner of a small mill: —

". . . Now as to prohibiting working under fourteen years of age, I think such a law would be very unjust and would break up many widowed families. My father died, leaving my mother with eight children from four to sixteen years old. We were poor people and had it not been that we worked in a cotton mill my mother would have had to divide her children and we would not have had the privilege of living with our mother and had a mother's care over us. As it was, we, or six of us, worked daily in the factory, and although it has been fifty-four years ago I am truly thankful there was such a place for us; for we made a living, owed nobody anything, and all grew up having a mother to watch over us, and although none of us ever amounted to much, we have never been considered bad people. Now to my knowledge there are many who were left like my mother was, and for our state to say to such ' Your children shall not work to make a living,' but have them put out, one here and another there, I have no language to tell you what I think of people who would try to control such things. . . . Now as to eleven hours a day. I don't think there should be a law to that effect. I favor it myself, but I do not know another's necessities, and I should not prevent him using his own judgment in making a living. Some can afford not to work at all, yet others cannot. The same commandment which says we must work six days, says we must rest the seventh. Now I don't know which is the greater sin, to be idle the six days or to work the seventh. Then if our law makers tell a part of us that we shall only work so much, I think they should say to the other parts that they *shall* work so much."

Another manufacturer, who has aided

churches and schools, established a library, and done much for the improvement of his operatives, wrote thus of the difficulty of excluding children, before the enactment of the labor law: —

"We do not want to work children under twelve years of age, — resist it all we can, — but when a mother brings an eleven-year-old boy to us, and pleads that school is over for the year, and she cannot look after her boy, that he roams the street contrary to her wishes, goes to the river with other boys, and she cannot keep up with him, and he wants to work in the mill, and she begs us to take him in as he is better off under control than out of it, what are we to do? The mother asked no wages for the boy but only wanted him where he could learn to work; we took the boy and he is earning fair wages."

The employers further claim, that since the rate per machine, or per unit of product, is the same regardless of the age of the operative, any charge of exploitation of children is unjust. It is true that the differences in individual wages in the same department are simply differences of skill, but the fact that the rate would probably rise with a curtailment of the labor supply is neglected.

The charge so often made, that the managers are tyrants who grind the helpless, cannot be sustained. They pay the market rate of wages, which is larger than the rate in agriculture, and the rate is steadily rising. They show in addition much personal interest in individuals and do many acts of kindness. Moreover, acts of cruelty would be injurious to the mills, as the neighboring mills will gladly advance transportation for a family of good operatives. The employers are not primarily to blame for the evils of child labor. Such labor is simply a stage in the development of an industrial society. The great numbers of families coming from the farms are the raw material out of which a skilled labor force is to be developed. Necessarily there are difficulties in adjustment.

The phenomenon is always present in an industrial transition. Hours were longer and conditions harder in England, in the *same stage* of industrial development. In New England, also, hours were longer. Mrs. Harriet H. Robinson, who went to work at Lowell when ten years of age for $2 a week, says: "The working hours of all the girls extended from five o'clock in the morning until seven at night, with one half hour for breakfast and for dinner. Even the doffers were forced to be on duty nearly fourteen hours a day. . . . I do not know why I did not think . . . of my work in the mill as drudgery. Perhaps it was because I *expected* to do my part towards helping my mother to get our living and had never heard her complain of the hardships of her life."[1]

Wonder has been often expressed that public opinion does not force the enactment of stringent labor legislation. So far there has been little attempt to organize this force. The growth of the industry has come too quickly. Regulation of farm work would seem absurd, and the realization of the difference between the work in the factory and on the farm comes slowly. Further, interference with the affairs

[1] Robinson, "Loom and Spindle" (1898).

of others is not yet popular, and anything smacking of class legislation meets with little favor, on general principles. The following expression of a farmer is typical of the attitude of mind of a large number, "I think that the less the state tends to supplant the family, the better." The agitation by outsiders has done more harm than good. The exaggerations have discounted the force of the whole argument, and have caused resentment of interference as well.

The beginning was made by the legislature of 1903, which passed the present act. It provides for a maximum week of sixty-six hours, and a minimum age of twelve. The parent must give a written statement of the ages of his children, and a false representation is punishable as a misdemeanor. The employer who knowingly employs a child under twelve is also liable to punishment. A stricter bill providing for raising the minimum age for girls and illiterate boys to fourteen, and absolutely forbidding night work by children, was defeated in 1905; but its passage will be urged again before the legislature of 1907.

Though the act of 1903 provided no system of inspection it has been generally obeyed, and the mill managers as a whole approve. Some, however, take the ground that their business is no more a proper matter for regulation than the farm or the sawmill. The operation of the law has sent many children into the schools, though not to the extent expected.

Agitation for compulsory school attendance is now in progress in connection with the movement for the increase of facilities for popular education. The great burden of illiteracy is leading many men to revise their belief regarding the proper limits of state interference with individual liberty. Organizations of women are beginning to advocate stricter laws for the protection of the children. A state Child Labor Committee has been organized (1906), and the creation of sentiment for stricter regulations will be attempted. Further legislation may be expected before many years.

## CHAPTER XIII

### THE NEGRO AS A COMPETITOR

SPEAKING broadly, in the South the right of the negro to earn a living by any sort of manual or mechanical labor has been recognized as a matter of course. Certain trades, as that of barber, have been almost monopolized by him. Contrary to conditions in the North, negro carpenters, bricklayers, plasterers, and plumbers work beside whites without question. On the farms, negroes and whites work together at all stages of the crops. Both are engaged in clearing forests and preparing wood or lumber for market. One wagon may be driven by a negro and another by a white. A white cobbler may share a shop with a negro. White men work with negroes in the tobacco factories, though usually in different processes.

But while public opinion accepts all this,

the working of negroes, particularly negro men, beside white women within walls would not be tolerated. Leaving any color prejudice out of consideration, the experience of the South with the "unspeakable crime" has been bitter. No association which might permit the possible lessening of the negro's deference toward white women would be allowed. It is a fixed belief, not susceptible to argument, that daily contact and association in the same work, under the same conditions, might tend to make the negro bolder and less respectful. For this reason the only negroes employed directly in the Southern textile industry are a few outside the mill proper, serving as laborers, draymen, firemen; and a smaller number engaged in some of the preparatory processes.

Heretofore the supply of white labor has been so abundant and so cheap that the adaptability of the negro to textile work has been a question chiefly of academic interest. A few farsighted mill men have realized that this supply of native white labor is not inexhaustible; but to test the capability of the negro would be an uncertain experiment which might cause the loss of time and capital, and even if successful seemed hardly worth the trouble and risk for the present. Some friends of the negro have hoped for the trial under favorable circumstances, as a means of discipline and development, but this interest has not been widespread. The general opinion expressed when the subject has been mentioned has been one of disbelief in its practicability, in spite of certain facts more or less well known.

While the industry was in the domestic stage, negro women on the large plantations spun yarn and wove cloth for plantation uses. Some of this work was well done, but no estimate of the number employed, nor of the value of the product, can now be made. Before the Civil War slave labor was employed in a few small cotton mills. The Rocky Mount Mill in Edgecombe County, North Carolina, employed negroes from 1820 to 1851. The following is taken from a private letter from the manager:—

"I took charge of the Rocky Mount mill in Nov. 1849. We worked at that time only negroes — nearly all of them slaves. There were 2 or 3 old issue free negroes. I introduced white labor in 1851. The whites seemed to think it *humiliating* to work in a cotton mill and I had much difficulty in getting them to go in. The mill was still making coarse yarns, 4's to 12's, put up in five pound bundles for the country trade — this was woven by country women on hand looms. When I could not sell the full product to the country merchants, the surplus was put in coarse filling for the Philadelphia market. I found the negroes to do pretty well on these coarse products, but the owners of the slaves began to object to their working in the mill and I substituted whites as soon as I could, and kept the mill going until destroyed by Federals in July, 1863."

Recently a few attempts to conduct mills entirely with negro labor have been made. The attempt at Charleston, South Carolina, was not successful, but the mill had previously failed with white labor. The old machinery was replaced by new and the mill was expected to pay dividends on this increased capitalization. The manager, Major J. H. Montgomery, was reported as attributing the failure principally to the location. In Charleston the bare necessities of subsistence are easily procured. Fish, oysters, vegetables, are cheap, the climate is mild, and little fuel and clothing give comfort. The usual attractions of the city were serious obstacles. Anything in the way of a pageant is exceedingly attractive to the negro, and it was difficult for these reasons to secure regular attendance. The manager expressed the belief that the experiment would have succeeded if it had been located in the country, away from the distractions of the town, where the operatives might have been better controlled, through living in factory tenements.

The Ashley and Bailey Company, silk manufacturers of Paterson, New Jersey, have established small silk mills at Fayetteville, North Carolina. The manager is a negro preacher, who has carefully selected his help by individuals and not by families. Before a youth is employed his parents must give the manager written permission to inflict corporal punishment if it is deemed necessary. Occasionally this permission is utilized, but the possession of the power has generally prevented the necessity of its exercise. The results of operation are said to be satisfactory, though the experiment has not yet gone far enough for a verdict, and the management refuses to give any information whatever upon the subject.

Another interesting experiment took place at Concord, North Carolina, a center of the Southern cotton mill industry. This was a cotton mill, not only operated, but owned and managed, by negroes. The moving spirit was a mulatto, Warren Coleman, who had an unusual career. He was born a slave and his reputed father was a white man, afterward distinguished by military and financial ability, who is said to have assisted the boy. Just after the Civil War, Coleman opened a little store, and succeeded through trading ability, industry, and economy. With his profits he began to buy cheap land on the outskirts of the town and erect cabins for negro tenants.

A house and lot costing from $125 to $300 could be rented for 50 cents to $1.25 a week. He built other houses with his rents, and at one time owned nearly 100. Valuable business sites were acquired, but these were not improved, as he hesitated to invest a large amount in a single venture. In 1900 his property was supposed to exceed $50,000.

Coleman conceived the idea that a cotton mill could be managed and operated by negroes, and began to agitate the matter. His motive was complex. The other mills in the town were almost phenomenally successful; his own past success as a financier had made him ambitious to be recognized as a factor in a broader field. Further, his race consciousness was strong and he desired to be considered the negro Moses, and to receive the applause gained by opening a new field of activity to his people.

The project was received with enthusiasm and every influence in the race was enlisted. Ministers recommended the enterprise from their pulpits; mass meetings making a strong appeal to race consciousness were held over the whole South, while the negro newspapers urged subscriptions as a duty to the race. The following extracts from negro papers will show the tone of race comment: —

"We know him personally, honest, enterprising, filled to overflowing with devotion to every movement wherein the negro's interest is fostered and promoted. He knows no failure. We know many other enterprises already fixed by Hon. W. C. Coleman, that are living monuments of glory to race as well as paying institutions."

— *Search Light* (Austin, Tex.), July 11, 1896.

"The greatness of the man appears particularly in the way he makes obstacles and difficulties, help and not hindrances. W. C. Coleman will rank with Abraham Lincoln as their practical friend and benefactor. One gave them freedom, the other will give them an industrial position."

— *Southern Age* (Atlanta, Ga.), Feb. 6, 1897.

"Let all colored men who have money to invest and race pride about them take stock in the mill."

— *Piedmont Indicator* (Spartanburg, S. C.), Dec. 12, 1896.

About $50,000 was subscribed, and the company was organized in 1897, with Coleman as Secretary and Treasurer. However, whether from jealousy or distrust few subscriptions were made by negroes living in Concord or the immediate vicinity.[1] Encouraged by the ready response, the capital stock was increased to $100,000, and subscriptions were sought from whites also. Those who responded were mill men, who were willing to risk a few dollars on the trial of the experiment, and a few philanthropists. A desirable tract of land was secured on the edge of the town, remote from the other mills, and building was begun.

When the collection of the subscriptions of the negroes began, difficulty ensued. It was found that laborers, washerwomen, etc., carried away by the enthusiasm of the moment, had subscribed amounts, the installments on which were as great as their total wages. Negro laborers and artisans had taken stock to be

[1] Testimony of J. P. Blackwell, bookkeeper, before the Industrial Commission, Vol. VIII.

paid in work on the buildings, but after a week or two, a certificate of stock in the future seemed less desirable than present cash payments, and the number of workers grew smaller. Much of this forfeited stock fell into Coleman's hands.

The work of construction dragged along and the building was not completed until 1901. Installments had been paid regularly on only a small part of the stock, and Coleman's holdings reached $12,000. Finally the building was finished, a few tenement houses were constructed, the railroad built a side track, and a mortgage was given for the equipment. Unfortunately Coleman, who had entire charge, seduced by apparent cheapness, put in secondhand English machinery, and the mill was handicapped from the beginning. A white superintendent from Easthampton, Massachusetts, was engaged, and operation was begun.

The time was unpropitious. The yarn market had not recovered from a period of demoralization, when many established mills with white operatives had run with a greatly reduced profit or with no profit at all. Operatives were to be made of individuals entirely unskilled and unused to any sort of regular mechanical work. The negro population of the town and surrounding country, however, was comparatively intelligent. The public schools were fairly efficient, and two schools maintained by Northern philanthropy had existed for years. Further, the negroes had been brought into close relations with the whites and had gained much from this contact.

The mill, nevertheless, did not pay, though the profit in yarns grew larger, and, in fact, did not run with any degree of regularity. A visit in 1902 found it shut down, — temporarily, the manager said, — but in reality little work had been done for weeks. The superintendent was absent; but his wife, who had practical knowledge of the business, assured me that the work was going on well, and that very skillful operatives were being produced. The operatives were regular and learned very rapidly, more rapidly than she had known whites to do in Massachusetts. She spoke approvingly of the conduct of the boys and girls, and declared that the negro overseers had been a success.

Through all her story, however, there seemed to run a note of insincerity. Her statements, both in sentiment and phrases, were too much like those of the manager who had urged me to visit the mill and mention it in the New York papers, saying frankly that he hoped a notice would bring him subscriptions to stock. The success of the mill had become a mania with him, and no opportunity to solicit subscriptions was lost. But he no longer spoke of it entirely as a business proposition, but asked aid on semi-philanthropic grounds. The same tendency had been observed earlier by the late Dr. Charles B. Spahr, who interviewed him in 1899.[1]

Coleman continued to furnish money for running expenses, sacrificing his real estate for

[1] "America's Working People," p. 48.

the purpose, until his resources were exhausted. In the fall of 1903 the management was turned over to a white merchant and cotton buyer of the town. This gentleman introduced several economies, engaged white overseers, and made the mill pay expenses until the high price of cotton in the spring of 1904 made further continuance impossible. Meanwhile Coleman died in April, 1904, and in June the mill was sold under the mortgage.

An examination of Coleman's affairs has shown that the mill owed him at least $12,000 which he had furnished at various times, though his books were in such confusion that the exact indebtedness could not be ascertained. Mr. White, who had charge of the mill during the last few months it was running, attributes the failure to the machinery, to inefficient management, and to a lack of working capital. Full production could not be secured from the worn machinery; but, by running slowly, the quality of the yarn produced was entirely satisfactory to the buyers, and regret was expressed at the discontinuance of operation.

Coleman had gained his property by economy and by investing his surplus in additional independent units of the same kind. With the rents from his houses, he built other houses. Close collections made him successful. In his store he kept only the staple groceries for which there was a steady demand. When greater problems were presented, he was not able to meet them. When profit or loss hinged upon the purchase of cotton on a certain day or a month afterward, or when accepting or rejecting a contract meant success or failure, his judgment was often at fault.

Further, his attitude toward his employees caused friction. The negro overseers were a failure. They were inclined to magnify their offices and to show favoritism. In the exercise of their power, they were sometimes over-lenient, but oftener overstrict, and docked the operatives on every opportunity. In this they were sustained and encouraged by Coleman, who seemed to consider every dime thus saved a real economy. Their overbearing disposition caused trouble, as it is proverbial that negroes will resent orders from one of their own color, which would be obeyed without question if they came from a white. The money needed for the operation of the mill was furnished in small sums when larger amounts would have been more economical. Neither cotton nor fuel had been supplied regularly. Often the mill was idle for hours waiting for a supply of cotton or coal. For lack of other fuel, the bagging from the cotton bales had been burnt, instead of being sold to the local ginners to be used again. Wages were paid irregularly, and the operatives were constantly changing.

Mr. White's verdict in regard to the labor is, on the whole, favorable. While a large number tested proved worthless, other women and girls were able to do efficient work, with the slow machinery, and developed also the quality of faithfulness and regularity. A few would be considered good average operatives in any Southern mill. In comparing general efficiency the wages paid must, however, be considered. While the white spinners in the town received 10 cents to 12½ cents the "side," it was easy to secure negroes at 5 or 6 cents. At this rate the best spinners made about $2.50 a week. The product per spindle was smaller, of course, than in the mills operated by white labor. The men employed were not so satisfactory as the women. Mr. White believes that with favoring circumstances a mill can be operated successfully with white overseers and negro operatives. However, he says that he would not attempt the experiment farther South where the negroes are perhaps less intelligent, nor in the vicinity of a city.[1]

It is clear, then, though both the Charleston and the Concord mills failed, that no verdict has been pronounced against negro operatives so long as low wages will draw them to the mills. There seems to be little about a mill which the

[1] Interview, 1904.

negro should not be able to learn. The processes are largely mechanical, and the difference between a good operative and a poor one is chiefly a question of care and dexterity. The negro is not by nature a machinist, but individuals can deal with machinery. The memory of a negro locomotive fireman who did the switching in the freight yards of a town where my boyhood was spent is vivid yet. Negro firemen and engineers of stationary plants are not uncommon. Of course many cannot comprehend even the elements of mechanics, but speaking broadly the difficulty with negro operatives is not an intellectual one.

The chief failings of all negro labor are temperamental and moral. The negroes as a class do not work except under direct compulsion. They do not like monotonous labor. They do not like to be alone nor to engage in any employment where they cannot communicate with their fellows. In the small Southern tobacco factories, the negroes talk and sing at their work as there is little machinery and no tension.

Whether enough negroes are to be found in a community who will keep up the monotonous routine of a cotton mill week after week, is the question to be solved. The negro was not long enough in slavery to make the willingness to work instinctive. He has not been long enough out of slavery to develop those ambitions which hold one to distasteful employment for the sake of ultimate satisfaction. Few have developed a pride in doing the given work as well as possible.

The negro dislikes to work regularly. The employers of domestic servants are necessarily liberal with "afternoons and evenings out." The employers of negro mechanics must allow numerous absences. Frequently a Northern born employer of negroes in the South, who attempts to enforce the same rules that he would where conditions of life are harder, fails entirely, when a Southerner who will endure more, succeeds, partially at least, in getting work done. To go to the yearly or semi-yearly circus, or to the campmeeting, sometimes lasting for a week or more, to attend a funeral arrayed in the gorgeous uniform of his lodge, are some of the negro's passions.

Perhaps the elevation of the negro's ideals of citizenship and of his standards of life will enable him successfully to enter the employments which the growing scarcity of white labor must soon open to him. Some negroes order their lives in accordance with the universal standards of good citizenship; but there is little pressure of public opinion among them on any question not directly connected with partisan politics, and their children too often revert to irresponsibility. The loafer stands as high as the laborer. Among the thoughtless his position is often higher, for he wears the cast-off clothing of the white man, and appears better than the laborer in overalls.

Procuring the means of a simple existence is too easy to make necessary the full employment of strength and time. Domestic servants seldom live on the premises, and demand the right to leave when the evening meal is over. Generally they consider all broken or left-over victuals their perquisite. A white family with a negro cook often supports from one to five colored persons, besides feeding any friend who comes to the kitchen on an errand or to visit. This fact helps to explain the number of loafers seen upon the streets of any Southern town. They are supported by the pilferings of a mother or sister, wife or sweetheart, and a few cents gained by holding a horse, carrying a note or a package, furnish tobacco and whisky. The white men for whom they do some little services turn over their discarded clothing, and too many desire little more.

Economic conditions are changing, however. In some sections white servants threaten to displace the negro. With the steadily rising price of food closer watch is kept on the pantry. The relations between the races are becoming more and more a matter of business, and the negro must work or become an habitual criminal. In view of the growing demand for labor, a stricter enforcement of the vagrancy laws does not seem unreasonable.

As was said above, the right of the negro to work has been unquestioned. During the operation of the Coleman mill there was not the slightest friction, and no prejudice was exhibited in the town toward the white overseers. Such a condition may not continue. If the negro holds his place in other industries, and enters the textile industry before a white operative class develops and becomes conscious of itself, extension of that employment is likely to cause a little jar. If he loses his industrial position, and, sometime in the future, after the mill operatives have become organized, attempts to enter, an intense race conflict may ensue. With the organization of the mill operatives, the relations between them and the operators will suffer a change. A conflict like that at Pana, Illinois, may follow the attempt to substitute colored for white labor, on account of a future strike or lockout.

# CHAPTER XIV

### CONCLUSIONS

We have now traced the development of a state from a collection of primitive frontier communities into one in which primitive conditions and somewhat advanced industrialism are strangely mingled. We have seen in the same neighborhood the oldest methods in agriculture and the most elaborate and costly machinery in manufacturing; the unskilled laborer and the expert operative.

A century ago the frugal population was almost self-sufficient, producing practically all that it consumed. The gradual decay of home manufacturing, and the increasing dependence upon other sections and other counties, have been shown. Then with the destruction and demoralization of the old system, we have seen a belated struggle for industrial position.

The simple country people who have always lived close to the soil have been drawn into the mills and factories, there to adjust themselves to a new environment. This process of adjustment naturally is not always easy. Necessarily it is often gained only after a considerable period, and then with pain and difficulty.

Such a period of friction is not peculiar to the section. All industrial transitions exhibit it to a greater or less extent. Perhaps because of the personal element in the relations with the employers it is less pronounced than usual. The tie between employer and employed is not at first a class relation, and the growth of the class idea has been slow.

The general conclusions which follow from the facts set forth in the text may be classified into those relating (1) to the industry itself; (2) to the employer; (3) to the operatives and their dependents; (4) to the state as a whole.

Though the discussion has not been concerned with the purely economic side of production, the position of the industry may be thus summarized: —

Mill buildings and tenements may be constructed much more cheaply than in New England. The cost of fuel is decidedly less. Those mills which procure their cotton from their immediate neighborhood save in freight charges; but the mills which must send to the Gulf states for their raw material are at a positive disadvantage. The freight on the cotton is often greater than the New England mill pays, and the freight on the product to the point of distribution is additional expense.

The labor cost has been less, due partly to lower money wages, partly to longer hours, and finally to the absence of strikes and other forms of industrial friction. At the same time the necessity of employing inefficient labor, or what amounts to the same thing, — a disproportionate amount of labor which has not attained average skill, — has increased the cost of production above the point which the lower rate of wages would indicate. That is, full production has not been secured from the machinery. Further, the rate of wages is rising and hours are being shortened.

Heretofore the mills have been engaged almost entirely upon coarse goods, but the tendency toward the finer grades is definitely marked. That the South, and North Carolina particularly, should gain the first place in the industry does not seem absurd. However, the industry is so strongly intrenched in New England, and the possibilities of foreign trade so immense, that the industry may continue to expand in both sections. If one section must lose, the South will survive, provided that skill in management is equal.

The manufacturers are not yet economic entrepreneurs. In most cases they were not trained in cotton mills, but entered the business after succeeding in something else. Some are shrewd and farsighted, few are harsh and despotic. Their success has been due more largely to general business experience, and to tact in the management of their employees, than to wide knowledge of the cotton business. At times it has been almost impossible to avoid making profits. Increasing competition will necessarily eliminate some of those now engaged in mill management.

With some detail and repetition that part of the rural population from which operatives come has been described. Their motives for coming have been analyzed, and their life around and in the mills has been discussed at length.

The attempt has been made to show the operatives as a whole and not a few unusual or abnormal examples. We have seen them to be honest, simple, and uneducated, but capable of development and training. Emphasis is laid upon the fact that they are neither degraded nor degenerate. In view of current misrepresentations, this fact cannot be stated too forcibly.

In regard to wages, the inevitable conclusion must be that, taking everything into consideration, the operatives are not wretchedly paid. While the wages are less than in New England, the demands made upon the wages are also less. With the increased reward of agricultural labor during the past five years, wages in the mills have risen decidedly. The pay is greater than in other local occupations open to those of no more training and skill. In fact the difference in favor of the factory is so great that only the natural inertia of a rural population combined with certain social disadvantages of factory labor prevents an oversupply.

Undoubtedly, a certain disrepute has, in the past, attached itself to factory labor in some localities. Perhaps the partial surrender of independence necessary has been responsible for some of this feeling. Then, too, around some mills moral conditions have not been beyond criticism.

A serious disadvantage from the standpoint of the student of social welfare is the tendency toward the destruction of family life. This is particularly true when the mill runs both night and day, and the family is divided. Further, where a definite part of the family income is directly attributable to a child, and that part is perhaps greater than the contribution of the parent, the natural relation of parent and child tends to be reversed.

While no defense of the employment of the child has been attempted or intended, the extent has been shown to be much less than has generally been supposed. Moreover, it would seem that some of the more serious phases of the problem belong to the transition period, and will correct themselves. The number of children employed grows less comparatively as the years pass.

In making comparisons with other sections in regard to hours of labor, employment of children, etc., it is only just to consider the suddenness with which manufacturing has been introduced into a society distinctly agricultural. Instead of comparing present conditions, it is fairer to compare North Carolina to-day with those sections when they were in the *same stage of industrial development*.

The problem of enriching the lives of these

people is still unsolved. The church is not holding its own, and no other social agency is taking its place. There is little around the factory village to develop the æsthetic and spiritual element. The daily life is, to a large extent, a round of toil, relieved only by physical pleasures. The large proportion of illiteracy, of course, increases the difficulty, and without compulsory school attendance a decrease will be slow. A comprehensive scheme of efficient agencies for social betterment remains to be developed.

The unusual relations between employer and employed heretofore existing have broken the shock between the life on the farm and at the mill. These relations, however, are passing away as the employer grows more "business-like," and the operative loses his rural habit of mind. A class consciousness is slowly developing among the workers, and the results will be momentous.

Whether future difficulties between the employer and employed will result in the introduction of negro labor into the mills, depends upon factors not purely economic. For a mill to discharge white operatives and introduce negroes would be a dangerous experiment from a social standpoint. With the increasing scarcity of white labor due to more prosperous conditions in other industries, a new mill might begin with negro operatives. The operatives must, however, be all white or all negro. In the present state of the public mind, indiscriminate employment is unthinkable. All these possibilities depend, however, upon the yet unproved capacity of the negro for such employment.

These tremendous problems of the industrial change have influenced the state as a whole. Yet since they have appeared gradually, some may deny any change. The student of social phenomena recognizes the decay of old ideals and the substitution of new. Political theories and prejudices, social customs and standards, ethical and religious values, are all affected. Nevertheless through all this confusion the influence of the old life unexpectedly persists, and strange inconsistencies appear. The state has not yet found itself; has not yet adjusted its agricultural philosophy to industrial conditions.

# APPENDIX A

## TABLE OF WEEKLY WAGES ACTUALLY PAID IN A NORTH CAROLINA TOWN, 1906

*Picker Room*

| | |
|---|---|
| Opener | $6.00 |
| Picker hand | 6.00 |
| Card hand | 6.00 |
| Boss carder | 12.00 |

*Spinning Room*

| | |
|---|---|
| Drawing frame hands | 6.00 |
| Slubber hands | 6.00 |
| Intermediate hands | 6.00 |
| Speeder hands | 6.00 |
| Spinners (12½c. to 15c. per side) | 3.00 |
| Doffers | 3.00 |
| Head Doffer | 3.60 |

| | |
|---|---|
| Spoolers | $4.50 to $6.00 |
| Twisters | 4.80 |
| Warper | 7.50 |
| Spinning overseer | 15.00 |
| Section hand | 7.50 |
| Twisting overseer | 7.50 |
| Band boy | 3.00 |
| Sweepers | 4.50 |
| Oiler and bander | 4.50 |

*Weaving Room*

| | |
|---|---|
| Filler | 4.50 |
| Creelers | 4.50 |
| Beam warper | 6.60 |
| Slasher tender | 7.50 |
| Drawing-in girls | 7.50 |
| Weavers ($3 to $9) | 5.40 |

*Finishing Room*

| | | | |
|---|---|---|---|
| Calendar | two | $6.00 | |
| Folder | men | 4.50 | |
| Baler | | | |

Weave boss . . $15.00
Section bosses . 9.00
Engineer . . . 9.00
Firemen . 6.00 to 7.50

# APPENDIX B

## Prices in Massachusetts and North Carolina

| ARTICLE | MASSACHUSETTS | | NORTH CAROLINA | |
|---|---|---|---|---|
| | 1897 | 1902 | 1897 | 1902 |
| Flour, Superfine, bbl. | $6.62½ | $5.20 | $6.00 | $5.00 |
| Flour, Family, bbl. | 5.80 | 4.69 | 5.00 | 4.50 |
| Meal, lb. | .03 | .03 | .01¼ | .02 |
| Rice, lb. | .07⅞ | .08 | .08¼ | .08¼ |
| Tea, lb. | .46⅞ | .54 | .50 | .50 |
| Coffee, Rio, lb. | .31¼ | .22 | .12¼ | .15 |
| Coffee, Roasted, lb. | .28 | .27 | .15 | .15 |
| Sugar, Coffee, lb. | .04¾ | .05¼ | .05 | .06¼ |
| Sugar, Gran., lb. | .05¾ | .05¼ | .06¼ | .07 |
| Molasses, N. O., gal. | .50 | .49¾ | .50 | .50 |
| Molasses, P. R., gal. | .49½ | .47 | .50 | .40 |
| Syrup, gal. | .52⅞ | .49¾ | .35 | .50 |
| Soap, lb. | .04¼ | .05¼ | .05 | .05 |
| Starch, lb. | .07½ | .08 | .07½ | .10 |
| **Meats** | | | | |
| Beef, Roast, lb. | .14¾ | .17½ | .08 | .10 |
| Beef, Soup, lb. | .05¾ | .07 | .05 | .05 |
| Beef, Steak, lb. | .25⅞ | .28 | .10 | .12½ |
| Veal, Fore Qrtr., lb. | .08 | .10¼ | .05 | .05 |
| Veal, Hind Qrtr., lb. | .12⅞ | .15¼ | .07 | .07 |
| Mutton, Fore Qrtr., lb. | .07½ | .10½ | .08 | .08¼ |
| Mutton, Leg, lb. | .11¾ | .16¼ | .10 | .10 |
| Mutton, Chops, lb. | .20 | .21¼ | .10 | .15 |
| Pork, Fresh, lb. | .10 | .14 | .05–.10 | .10 |
| Pork, Salted, lb. | .09⅞ | .12¼ | .07½ | .10 |
| Hams, lb. | .13¼ | .13¾ | .12 | .14 |
| Shoulders, lb. | .09 | .10 | .11 | .13 |
| Sausage, lb. | .10⅞ | .12¼ | .10 | .10 |
| Lard, lb. | .08 | .13¼ | .08¼ | .10 |
| Butter, lb. | .24¼ | .30¼ | .15 | .18 |
| Cheese, lb. | .14 | .16 | .17½ | .20 |

## APPENDIX B (*Continued*)

**Prices in Massachusetts and North Carolina**

| Article | Massachusetts | | North Carolina | |
|---|---|---|---|---|
| | 1897 | 1902 | 1897 | 1902 |
| **Vegetables** | | | | |
| Potatoes, White, bu. | $1.01⅓ | $1.14⅓ | $0.70 | $0.75 |
| Potatoes, Sweet, bu. | | | .60 | .50 |
| Milk, qt. | .05⅜ | .06¼ | .05 | .05 |
| Eggs, doz. | .23½ | .21½ | .12½ | .12½ |
| Board, Men, per week | 4.62 | 3.91 | 1.50 | 1.50 |
| Board, Women, per week | 3.66 | 3.34 | 1.50 | 1.50 |
| **Fuel** | | | | |
| Coal, Hard, ton | 6.00 | 6.65¾ | 4.50 | 4.50 |
| Wood, Hard, cord | 8.41⅛ | 8.25 | 1.60 | 1.75 |
| Wood, Pine, cord | 6.97 | 6.79¼ | 1.50 | 1.60 |
| **Dry Goods** | | | | |
| Shirting, 4–4 Brown, yd. | .08½ | .06¾ | .05½ | .07½ |
| Shirting, 4–4 Bleached, yd. | .08½ | .08¾ | .06 | .10 |
| Sheeting, 9–8 Brown, yd. / Sheeting, 9–8 Bleached, yd. | .09¾ | .16 | .09 | .10 |
| Cotton Flannel, yd. | .10 | .10¾ | .09 | .10 |
| Ticking, yd. | .11 | .13½ | .12½ | .12½ |
| Prints, yd. | .05⅛ | .06 | .05 | .06¼ |
| Shoes, Men's | 2.05½ | 1.99½ | 1.50 | 1.50 |
| **Rent** | | | | |
| 4-Room Tenements | 8.63 5/7 | 12.14 | 3.00 | 4.00 |
| 6-Room Tenements | 11.61 | 19.30 | 4.00 | 6.00 |

# APPENDIX C

## Comparison of Prices of Selected Commodities in Similar Towns in Connecticut and North Carolina, April, 1904

| Article | Conn. | N. C. | Article | Conn. | N. C. |
|---|---|---|---|---|---|
| Flour, bbl. | $6.25 | $5.50 / 6.50 | Hams, lb. | $0.15 | $0.15 |
|  |  |  | Shoulders, lb. | .11 | .11 |
| Meal, lb. | .05 | .02 | Sausage, lb. | .14 | .12½ |
| Rice, lb. | .07½ | .08¼ | Lard, lb. | .10½ | .10 |
| Tea, lb. | .50 | .50 |  |  |  |
| Coffee, lb. | .30 | .15 | **Fuel** |  |  |
| Sugar, Coffee, lb. | .05 | .06¼ | Coal, Soft, ton | 5.50 | 3.75 |
| Sugar, Gran., lb. | 18 lb. for $1.00 | 16 lb. for $1.00 | Wood, Hard, cord | 10.00 | 2.00 |
|  |  |  | Wood, Soft, cord | 7.00 | 1.75 |
| Molasses, N. O., gal. | .55 | .50 | **Dry Goods** |  |  |
| Molasses, P. R., gal. | .43 | .40 | Shirting, |  |  |
| Syrup, gal. | .55 | .50 | Unbleached, yd. | .12½ | .10 |
| Butter, lb. | .22 | .15 | Shirting, |  |  |
| Cheese, lb. | .15½ | .20 | Bleached, yd. | .15 | .12 |
| Milk, qt. | .06 | .05 | Sheeting, |  |  |
|  |  |  | Unbleached, yd. | .08 | .10 |
| Eggs, doz. | .31 | .12½ to .15 | Sheeting, |  |  |
|  |  |  | Bleached, yd. | .10 | .12 |
| Potatoes, bu. | .95 | 1.00 | Cotton Flannel, yd. | .08 | .08¼ |
| **Meats** |  |  | Prints, yd. | .07 | .06¼ |
| Beef, Roast, lb. | .16 | .10 | Shoes | 1.75 | 1.50 |
| Beef, Soup, lb. | .08 | .05 |  |  |  |
| Beef, Steak, lb. | .18 | .12½ | **Rent** |  |  |
| Veal, Fore Qrtr., lb. | .11 | .06 | 4-Room Tenements, wk. | 1.10 | 1.00 |
| Veal, Hind Qrtr., lb. | .20 | .08 |  |  |  |
| Mutton, Fore Qrtr., lb. | .10 | .10 | 6-Room Tenements, wk. | 2.00 | 1.50 |
| Mutton, Leg, lb. | .16 | .12½ | Board and Lodging | 2.00 to 3.00 | 1.50 |
| Pork, Fresh, lb. | .14 | .10 |  |  |  |
| Pork, Salt, lb. | .14 | .10 |  |  |  |

# APPENDIX D

## Weekly Wages paid in Seven North Carolina Mills, 1904

| Occupation | Rate per Week | Occupation | Rate per Week |
|---|---|---|---|
| **Picker Room** | | **Weaving Room** | |
| Opener | $4.50 | Filler | $3.90 |
| Picker Hand | 5.10 | Creelers | 4.00 |
| Card Hand | 4.50 | Beam Warper | 4.50 |
| Boss Carder | 12.00 | Slasher Tender | 6.00 |
| | | Drawing-in Girls | 6.00 |
| **Spinning Room** | | Weavers, $2.50 to 9 | 5.40 [1] |
| Drawing Frame | 4.50 | | |
| Slubber Hands | 5.40 | **Finishing Room** | |
| Intermediate Hands | 5.40 | Calendar ⎫ | ⎧ 6.00 |
| Speeder Hands | 4.50 | Folder  ⎬ 2 men | ⎨ |
| Spinners, $1.20 to $6 | 3.00 [1] | Baler  ⎭ | ⎩ 4.50 |
| Doffer, Head | 3.60 | Weave Boss | 15.00 |
| Doffers | 2.40 | Section Bosses | 8.40 |
| Spoolers | 4.00 | | ⎧ 7.50 |
| Twisters | 4.80 | Engineer | ⎨ to |
| Warpers | 7.50 | | ⎩ 9.00 |
| Overseer of Spinning | 10.50 | Firemen | 6.00 |
| Section Hand | 7.00 | | |
| Overseer of Twisting | 7.00 | | |
| Band Boys | 2.50 | | |
| Sweepers | 3.60 | | |
| Oiler and Bander | 3.60 | | |

[1] On account of variations in number and skill of these operatives the exact average wage is seldom the same for two successive weeks.